Alternating Current Field Measurement Technique for Detection and Measurement of Cracks in Structures

Wei Li · Xin'an Yuan · Jianming Zhao ·
Xiaokang Yin · Xiao Li

Alternating Current Field Measurement Technique for Detection and Measurement of Cracks in Structures

 Springer

Wei Li
Department of Mechanical and Electrical
Engineering
China University of Petroleum (East China)
Qingdao, Shandong, China

Xin'an Yuan
Department of Mechanical and Electrical
Engineering
China University of Petroleum (East China)
Qingdao, Shandong, China

Jianming Zhao
Department of Mechanical and Electrical
Engineering
China University of Petroleum (East China)
Qingdao, Shandong, China

Xiaokang Yin
Department of Mechanical and Electrical
Engineering
China University of Petroleum (East China)
Qingdao, Shandong, China

Xiao Li
Department of Mechanical and Electrical
Engineering
China University of Petroleum (East China)
Qingdao, Shandong, China

ISBN 978-981-97-7254-4   ISBN 978-981-97-7255-1   (eBook)
https://doi.org/10.1007/978-981-97-7255-1

This work was supported by China University of Petroleum (East China).

This Springer imprint is published by the registered company Springer Nature Singapore Pte Ltd.
The registered company address is: 152 Beach Road, #21-01/04 Gateway East, Singapore 189721, Singapore

If disposing of this product, please recycle the paper.

Preface

The nondestructive testing (NDT) method is an effective means to find and evaluate defect, which provides support for the safety pre-warning and the maintenance decision of structure in the industrial field. In the 1980s, there were urgent requirements for an NDT technology for inspection and evaluation of fatigue cracks at welded intersections in the offshore underwater structures in North Sea. The conventional NDT methods were not practicable in this special underwater environment and quantitative evaluation requirements. Thus, the alternating current field measurement (ACFM) technology was developed originally to inspect the underwater structures by the researchers in mechanical engineering department at University College London. In the past 40 years, the theory model, inspection method and equipment of ACFM have achieved rapid progress for the advantages of the noncontact capability, the reduction of lift-off effect and the size, which are used widely in the ocean engineering, power industry, rail traffic and special equipment field. In the past years, colleagues mainly focused on the study of basic theory, probe and system design and structure inspection. This book is divided into three parts. In the first part, three chapters are employed to introduce the basic theory of ACFM, which lays the theoretical foundation for the following parts. In the second part, three chapters are introduced to introduce the design and testing of ACFM probe, instrument and software, which provide guidance and reference for the development of the ACFM instrument. In the third part, three chapters are introduced to explain the visualization research in ACFM, which provide guidance for technical engineering application. ·

Qingdao, China   Wei Li
ronald8044@163.com

Xin'an Yuan
xinanupc@163.com

Jianming Zhao
jianmingzhao123@163.com

Xiaokang Yin
xiaokang.yin@hotmail.com

Xiao Li
lix2020@upc.edu.cn

Contents

Research on Real-time and High-Precision Cracks Inversion Algorithm for ACFM Based on GA-BP Neural Network

1 Introduction

Alternating current field measurement (ACFM) technology was initially applied to weld inspection on offshore drilling platforms. Due to its characteristics of non-contact measurement, no calibration required, low lift-off effect, and accurate mathematical model, it has been widely applied in fields such as petrochemicals, railway transportation, nuclear energy, etc. [1–3]. The principle of ACFM technology is shown in Fig. 1, where the excitation probe induces a uniform current on the surface of the workpiece. When the current passes through the defect vertically, it will bypass the ends and bottom of the defect, causing distortion in the surrounding magnetic field. The X-direction magnetic flux density Bx produces a trough, and the depth of the trough can reflect the depth of the defect. Similarly, the Z-direction magnetic flux density Bz generates peaks and troughs, and the distance between them can indicate the length of the defect [4, 5].

In the field of non-destructive testing, obtaining characteristic signals from defects is referred to as the forward model, while obtaining defect shapes from characteristic signals is referred to as the inverse model [6]. It is generally easier to obtain characteristic signals from defects, while it is more complex to invert defect shapes from characteristic signals. However, the information of defect shape can reflect the degree of structural damage, which is a matter of great concern in structural assessment [7]. The ACFM technique has an accurate mathematical model, where the characteristic signal Bz reflects the crack length information and Bx contains the crack depth information, providing good conditions for defect profile inversion (length and depth). Traditional ACFM interpolation inversion algorithms are relatively simple, but they have low accuracy and poor real-time performance [8, 9]. The inversion algorithms that utilize characteristic signals have high requirements for the sample database, and the mutual influence between characteristic signals makes the inversion process complex. It is necessary to analyze the characteristic signals or

© The Author(s) 2025
W. Li et al., *Alternating Current Field Measurement Technique for Detection and Measurement of Cracks in Structures*, https://doi.org/10.1007/978-981-97-7255-1_1

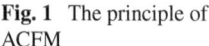

Fig. 1 The principle of ACFM

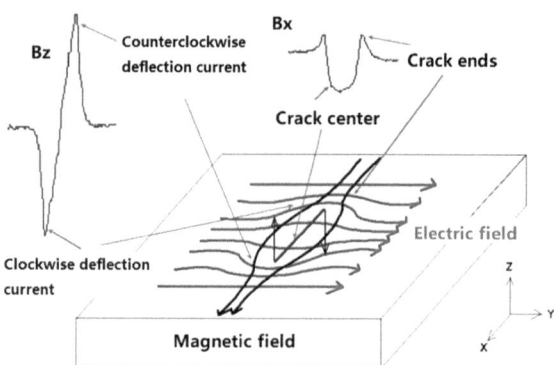

determine the butterfly chart after completing the defect scanning, making real-time determination and inversion difficult to achieve [10].

To solve the above problems, the author established the electromagnetic coupling simulation model of ACFM probe, obtained the sample database of crack inversion characteristic signals in real time through the energy spectrum and phase threshold determination method, and introduced Back Propagation, BP neural network and Genetic Algorithm (GA) are used to retrieve ACFM crack length and depth with high precision. In this paper, the GA-BP neural network based ACFM real-time high-precision inversion algorithm provides a new method for real-time crack inversion.

2   The FEM Model of Electromagnetic Coupling ACFM Probe

Select the 3-D circuit coupling stranded coil CIRCU124 in ANSYS, set its options as needed to make it a power supply, and generate a coupling unit with the K node and the node at the position where the probe model needs to be combined with the E command. Since the current in the coil and the potential drop through the coil terminal are unique, the degrees of freedom of coil current per turn (CURR) and potential drop (EMF) degrees of freedom through the coil terminal are separately coupled together. A 3D ACFM probe electromagnetic coupling motion simulation model was established, as shown in Fig. 2. In the model, the excitation probe is a U-type current-carrying coil [11, 12]. Under the coil, X coil (to extract the induced voltage value of the X direction magnetic field) and Z coil (to extract the induced voltage value of the Z direction magnetic field) are provided. In the simulation model, the crack length is 15 mm, the width is 0.8 mm, the depth is 5 mm, the loading excitation voltage is 3 V, the frequency is 6 K Hz, and the lifting height of the detection coil is 2 mm.

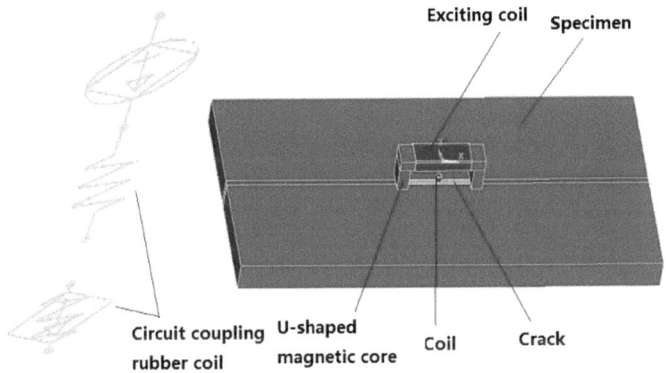

Fig. 2 The FEM model of electromagnetic coupling ACFM probe

A harmonic solver was selected, a path was established along the top of the crack, and the voltage Ex and Ez in the coils in the X and Z directions above the path were extracted respectively, as shown in Fig. 3. Ex maintains a certain value in the non-cracked region (the background magnetic field is in the X direction), and when it enters the cracked region (14–29 mm), the magnetic flux density in the X direction decreases, resulting in a trough in Ex. Because the extracted induced electromotive forces are all positive, Ez is basically 0 in the region without entering the crack, and two peaks appear in the aggregation area at both ends of the crack (14 and 29 mm). The Ex and Ez rules are consistent with the ACFM principle, which shows that the simulation model of the electromagnetic coupling ACFM probe established in this paper is correct.

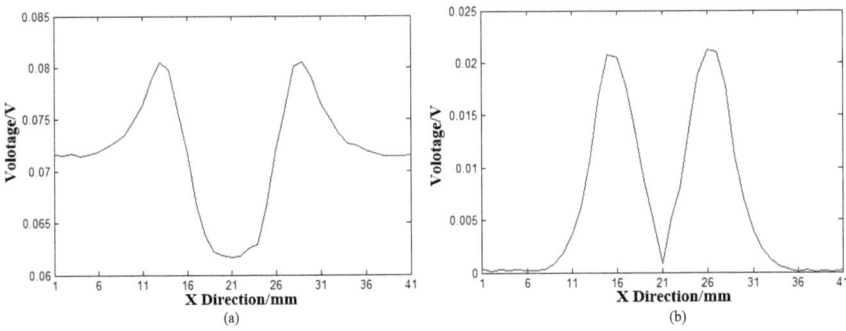

Fig. 3 The simulation results of crack characteristic signals. **a** Bx. **b** Bz

3 Real-Time Determination Method for Defect Characteristic Signals

Traditional ACFM technology often adopts the feature recognition method, that is, according to the features of Ex and Ez or the butterfly diagram composed of Ex and Ez [13]. Because the identification method based on characteristic signal or butterfly diagram needs to distinguish according to the characteristics of the obtained signal after the probe has completely swept the crack, it is difficult to realize the real-time data storage and processing of the defect characteristic signal. In this paper, the energy spectrum and phase threshold discriminant methods can be used to save the effective signal characteristic values of defects in real time according to the threshold, which lays a foundation for real-time and high-precision defect inversion.

As shown in Fig. 3, since the distance between the peaks of Ex (13–29 mm) is larger than the peak-to-peak distance of Ez (14–28 mm) and larger than the crack length, Ex contains more defect information (depth and length, especially the safety margin for length estimation). In this paper, the electromotive force Ex of the X coil is converted into the energy spectrum by the square of the Fourier transform, as shown in Fig. 4. Within the peak distance of Ex (dashed line range), the maximum value of the energy spectrum is taken and a horizontal line is numerically drawn. The intersection position of the horizontal line and the Ex energy spectrum (solid line region) is obviously greater than the Ex peak value, then the energy spectrum information located below contains all the length and depth information of the crack. Therefore, the peak value of Ex energy spectrum is selected as the threshold for real-time crack determination.

At the same time, because the direction of the magnetic flux density inside the crack zone changes when the detection coil enters the crack zone, the phase of Ez will change when the Z coil enters the defect. As shown in Fig. 5a, Ez phase has a sudden change at the crack region (14–29 mm). Ez phase derivative is shown in Fig. 5b, and it can be seen that Ez phase derivative has an obvious peak value in the crack region. Therefore, the derivative of Ez phase is selected as the basis for real-time crack determination. The crack characteristic signal can be obtained in real time according to the Ex energy spectrum and Ez phase threshold real-time determination method. The real-time determination method designed for crack characteristic signal is shown in Fig. 6. The main steps are as follows:

(1) Perform phase-lock amplification and data processing on the collected Ex, Ez and excitation signals to obtain the Ex energy spectrum and Ez phase.

(2) Determine the relationship with the threshold size. If the value is less than or equal to the value, the data is saved. Take the derivative of Ez phase to get Dz.

(3) Determine whether Dz meets the requirements (there is a margin for phase judgment in this paper, and the threshold of Ez phase derivative is set to 180). If Dz is greater than or equal to 180, it is regarded as valid data, and the saved data is regarded as the defect characteristic signal sample database. If Dz is less

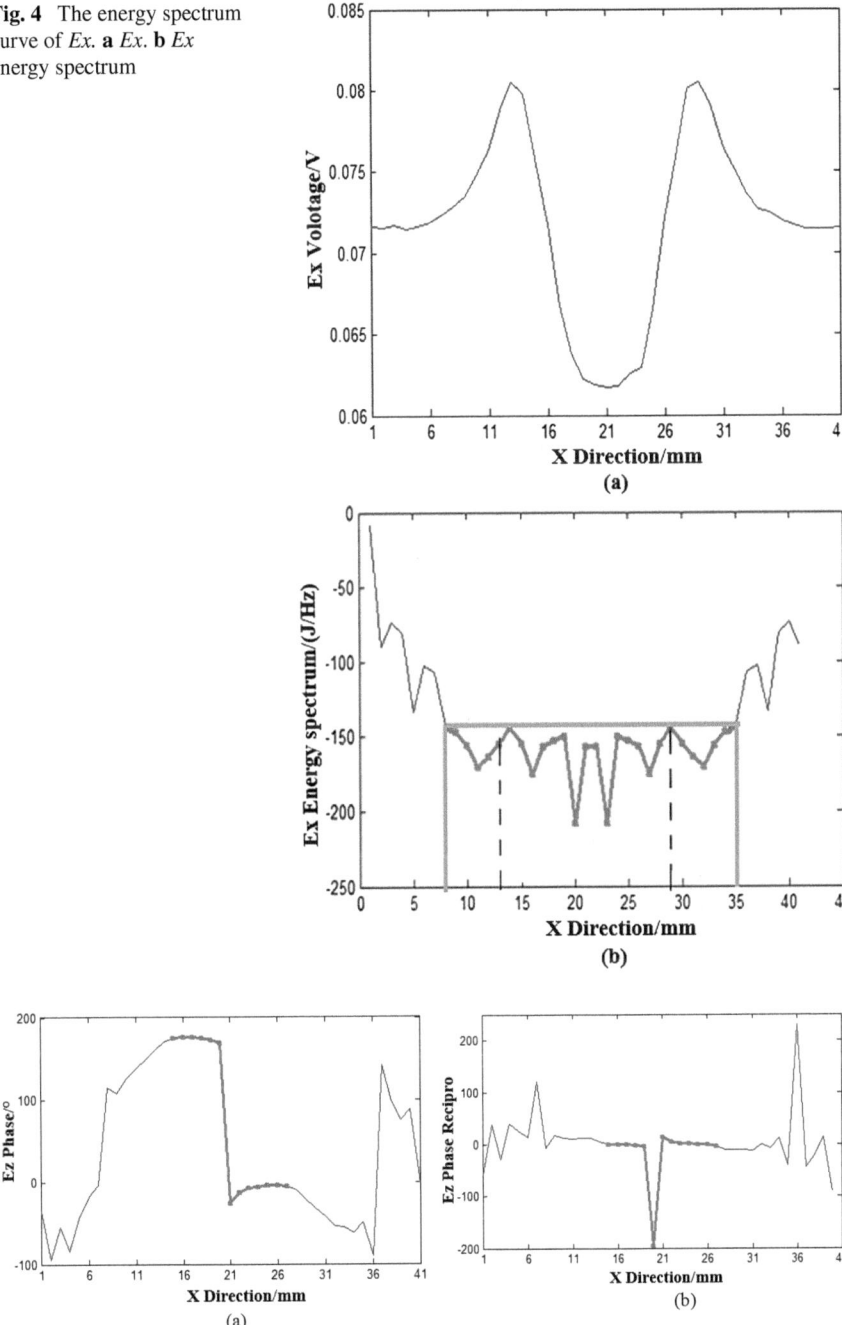

Fig. 4 The energy spectrum curve of *Ex*. **a** *Ex*. **b** *Ex* energy spectrum

Fig. 5 The phase of *Ez*. **a** Different position phase of *Ez*. **b** Pase reciprocal of *Ez*

Fig. 6 The defect real-time
determination method

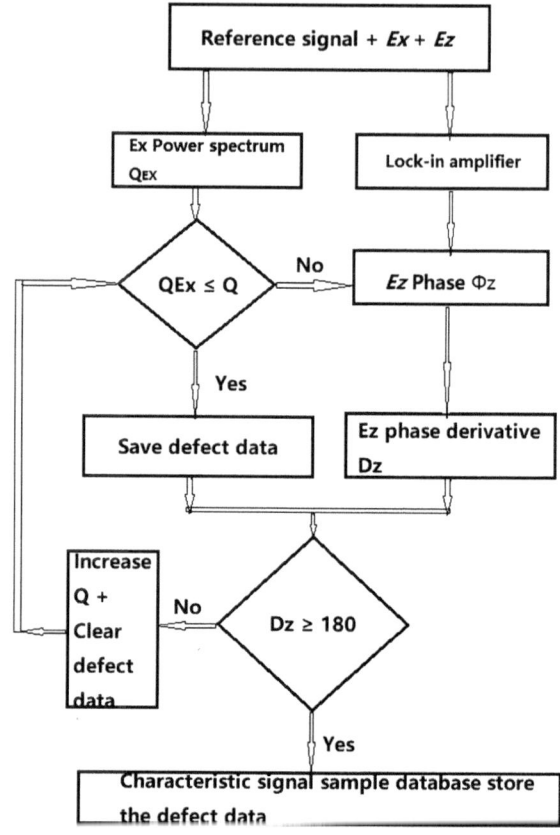

than 180, increase the threshold, clear the saved defect data, and continue to determine the threshold size. If the value is greater than the value, the data is discarded.

By updating the threshold value in real time, the algorithm can effectively avoid the misjudgment of the crack characteristic signal, and effectively save the crack characteristic signal in time when the detection coil enters the crack zone, so as to achieve the purpose of real-time crack determination without completely sweeping the crack.

4 Real-Time Crack Inversion Experiment

4.1 Detection System Design

According to the principle of ACFM technology, the crack real-time inversion experimental system is designed, as shown in Fig. 7a. In this paper, the ACFM excitation
probe uses U-type current carrier coil, and the detection coil uses coil sensor. The
signal generator generates a sinusoidal signal of amplitude 1 V and frequency 6 kHz,
which is transmitted to the excitation coil through power amplification. The exciting
coil generates a uniform exciting current region on the specimen surface, and the
space magnetic field distortion is caused by the current passing through the defect.
The detection coil picks up the Bx and Bz information and converts it into electrical signals Ex and Ez. Ex and Ez are transmitted from the acquisition system to
the computer through conditioning circuits (amplification and filtering). The defect
characteristic signal can be obtained in real time by using Ex energy spectrum and Ez
phase threshold method. The real crack real-time inversion detection system finally
built is shown in Fig. 7b.

The specimens in this paper are mild steel plates, on which rectangular artificial
cracks of different sizes are carved by EDM technology. A PLC controlled scanning
platform was used to drive the ACFM probe (excitation and detection coil) along one
of the cracks (length 40 mm and depth 6 mm) for uniform scanning. The measurement
and control software collected the position information of Bx, Bz and the probe, and
the defect characteristic signals obtained were shown in Fig. 8.

In order to realize real-time and high-precision inversion quantification of defects,
sensitivity is introduced as a characteristic quantity of defect size inversion [14]:

$$S_x = \frac{M_x}{E_{x0}} = \frac{E_{x0} - E_{x\min}}{E_{x0}}, S_z = \frac{E_{z\max}}{E_{x0}}$$

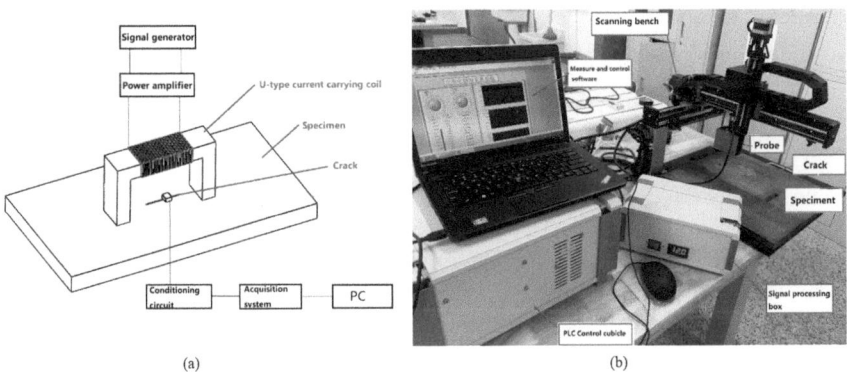

(a)   (b)

Fig. 7 The real-time and high- precision cracks inversion system for ACFM. **a** System schematic
diagram. **b** Physical inspection system

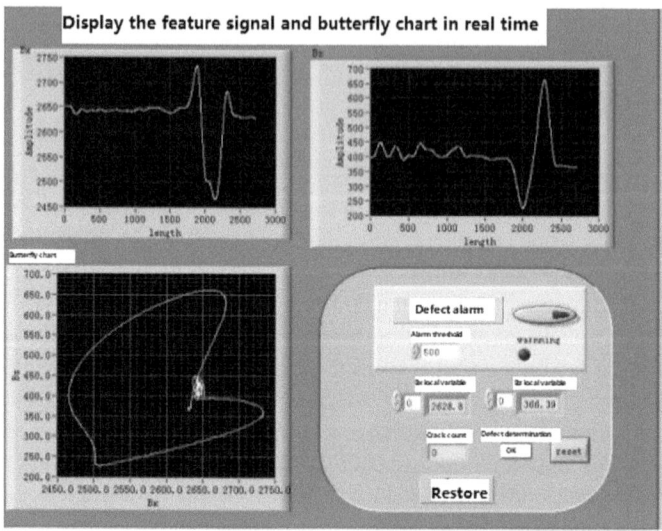

Fig. 8 The characteristic signals of crack

L_Z: E_Z: peak-to-peak interval of component waveform; E_{X0}: component signal amplitude away from the crack; $E_{X\,min}$: minimum signal amplitude; $E_{Z\,max}$: indicates the maximum value of signal distortion.

By analyzing the crack characteristic signals obtained by software in real time, the results show that Ex sensitivity Sx is 9.2%, Ez sensitivity Sz is 26.63%, and the distance between Fz peaks is 38.46 mm. At the same time, cracks of different depths and lengths were detected, and the signal characteristic quantities of different crack sizes were obtained, as shown in Table 1.

4.2 Defect Inversion Algorithm Based on GA-BP

As shown in Fig. 9, this paper uses the newff function in MATLAB to create a 4-layer BP neural network. By default, the default initnw is used to initialize weight and bias [15]. Although the initnw method enables the active regions of neurons in each layer to be roughly flat distributed in the input space, the network trained by this method has the problem of network structure uncertainty, that is, the initial weights and bias values are random. Therefore, BP algorithm has some characteristics, such as slow convergence speed, unguaranteed convergence to the global minimum point, and difficult to determine the network structure [16].

Genetic algorithm, based on biological evolution, has good convergence and robustness, and the calculation time is less when the calculation accuracy is high [17, 18]. In this paper, a step genetic algorithm is added to optimize the weight and bias of the neural network (GA-BP neural network model) before the training of BP

Table 1 The characteristic signals database of different crack size

S. No.	E_X sensitivity S_X/%	E_Z sensitivity S_Z/%	E_Z peak-to-peak interval L_Z/mm	Defect depth D/mm	Defect length L/mm
1	39.31	41.44	8.29	8	10
2	31.65	38.69	18.85	8	20
3	25.52	35.64	2861	8	30
4	10.97	33.87	38.82	8	40
5	31.68	39.31	8.33	6	10
6	24.62	35.74	18.69	6	20
7	17.84	31.58	28.53	6	30
8	9.20	26.63	38.46	6	40
9	23.42	33.24	8.45	4	10
10	17.38	26.76	18.62	4	20
11	10.92	23.11	28.72	4	30
12	7.58	18.87	38.69	4	40
13	14.61	12.47	8.46	2	10
14	10.30	14.23	18.52	2	20
15	7.10	12.59	28.36	2	30
16	6.28	9.91	38.75	2	40
17	9.87	23.12	38.73	5	43
18	14.23	24.85	23.76	4	25
19	27.95	37.43	13.14	6	15

Fig. 9 The model of layer BP neural network

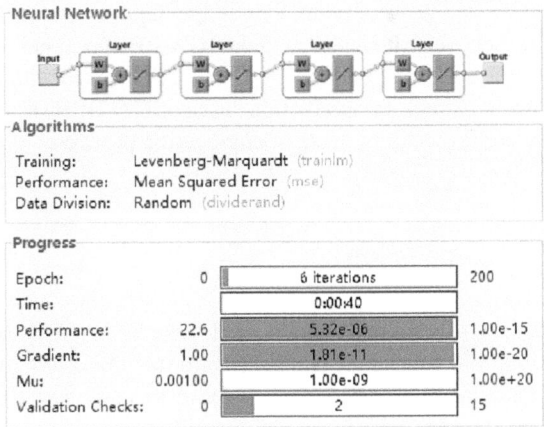

network, so as to determine an optimal set of initial weights and bias, so as to optimize the structure of BP artificial neural network [19–21]. The main program instruction of BP neural network based on genetic algorithm initialization is as follows:

[net, IW, IB, W1, W2, B1, B2]
= fgaBP(XX, YY, eranum, popsize, pCross, pMutation, pInversion, options).

In the GA-BP neural network for defect inversion in this paper, XX input vector includes three characteristic quantities of sensitivity and sensitivity as well as pek-valley distance of component waveform, while YY output vector is two characteristic quantities of defect length and depth. The neuron matrix of the four-layer BP network model established by this method is [2, 5, 6]. Groups 1–16 in Table 3 are taken as training samples and groups 17–19 as test samples. Since the value range of defect size is also quite different, in order to facilitate the convergence of BP-GA network and improve the prediction accuracy, the sample data in Table 3 should be normalized within the interval [− 1, 1] before training. The weight and bias process of BP neural network solved by genetic algorithm is shown in Fig. 10a, and the training process of BP neural network is shown in Fig. 10b.

The GA-BP neural network established above was learned, and the cracks of group 17 (length 43 mm and depth 5 mm) in the sample database were predicted. The crack prediction results were shown in Fig. 11. The inversion results show that the crack size of group 17 is 40.70 in length and 4.72 in depth, with length error of 5.35% and depth error of 5.60%.

Similarly, GA-BP neural network was used to invert the cracks in 18 groups and 19 groups of samples, and the inversion results were shown in Table 2. The inversion error of crack length and depth in group 18 is 6.12% and 7.00% respectively. The inversion error of crack length and depth of group 19 samples is 9.40% and 6.51% respectively. It can be seen that GA-BP neural network has better prediction ability for deeper cracks and longer cracks. Compared with the actual crack size, the relative error of crack size prediction based on GA-BP neural network is less than 10%, which meets the requirements of engineering practice.

5 Conclusion

In this paper, the electromagnetic coupling ACFM simulation model was established with the help of ANSYS software, the characteristic crack signal was obtained in real time by using the energy spectrum and phase threshold determination methods, the ACFM crack inversion experimental system was built and the crack detection experiment was carried out, and the GA-BP neural network was used to achieve real-time and high-precision real-time inversion of crack size. The main conclusions were as follows:

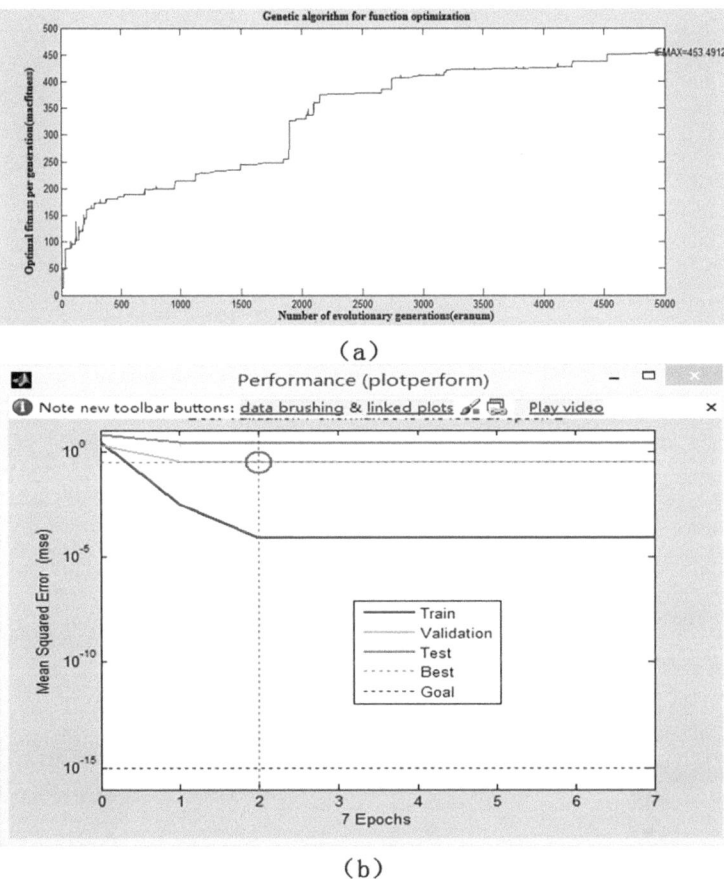

(a)

(b)

Fig. 10 The training process of GA-BP. **a** Genetic algorithm optimization process. **b** BP Neural network training process

(1) The electromagnetic coupling ACFM simulation model established in this paper can accurately obtain the voltage E_x and E_z inside the X coil and Z coil above the crack.

(2) The threshold determination method based on E_x energy spectrum and E_z phase derivative can obtain the crack characteristic signal in real time.

(3) The crack inversion algorithm based on GA-BP neural network can effectively predict the length and depth of defects, and the prediction error is less than 10%, which meets the requirements of engineering practice.

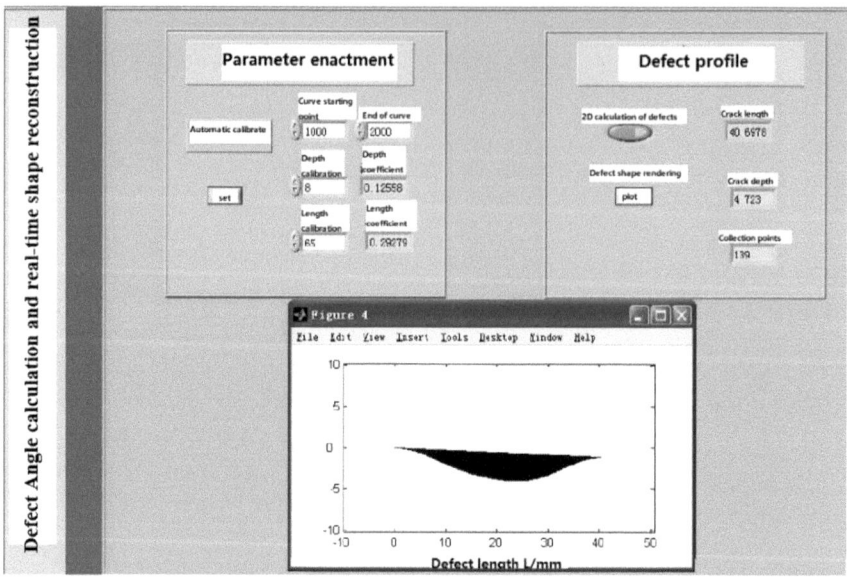

Fig. 11 The test results of GA-BP in 19th data

Table 2 The forecasting result of test sample based on GA-BP neural network

S. No.	Depth inverse normalization/mm	Relative error/%	Length inverse normalization/mm	Length inverse normalization/%
17	4.72	5.60	40.70	5.35
18	3.72	7.00	23.47	6.12
19	6.51	8.50	13.59	9.40

References

1. Lugg MC (2012) The first 20 years of the A.C. field measurement technique. In: 18th world conference on non-destructive testing (WCNDT), South Africa, pp 16–20
2. Li W, Yuan XA, Chen GM et al (2015) Research on in-service detection for axial cracks on drill pipe using the feed-through alternating current field measurement. J Mech Eng 51(12):8–15
3. Nicholson GL, Davis CL (2012) Modelling of the response of an ACFM sensor to rail and rail wheel RCF cracks. NDT&E Int 46:107–114
4. Li W, Yuan XA, Chen GM et al (2014) A feed-through ACFM probe with sensor array for pipe string cracks inspection. NDT&E Int 67:17–23
5. Qi YL, Chen GM, Zhang YT (2004) Numerical simulation on alternating current field measurement and sensitivity analysis of detected signal. J Univ Pet China (Edn Nat Sci) 28(3):65–68
6. Ramuhalli P, Udpa L, Udpa SS (2002) Electromagnetic NDE signal inversion by function-approximation neural networks. IEEE Trans Magn 38(6):3633–3642
7. Maazi M, Benzaim O, Glay D et al (2008) Detection and characterization of buried macroscopic cracks inside dielectric materials by microwave techniques and artificial neural networks. IEEE Trans Instrum Meas 57(12):2816–2819

8. Li W (2007) Research on ACFM based defect intelligent recognition and visualization technique. China University of Petrol, Dongying
9. Li W, Chen GM (2009) Defect Visualization for alternating current field measurement based on the double U-shape inducer array. J Mech Eng 45(9):233–237
10. Noroozi A, Hasanzadeh RPR, Ravan M (2013) A fuzzy learning approach for identification of arbitrary crack profiles using ACFM technique. IEEE Trans Magn 49(9):5016–5027
11. Li W, Chen GM, Yin XK et al (2013) Analysis of the lift-off effect of a U-shaped ACFM system. NDT&E Int 53:31–35
12. Li W, Chen GM, Li WY et al (2011) Analysis of the inducing frequency of a U-shaped ACFM system. NDT&E Int 44:324–328
13. Lugg M, Topp D (2006) Recent developments and applications of the ACFM inspection method and ACSM stress measurement method. In: Proceedings of ECNDT, Berlin, Germany
14. Nicholson GL, Kostryzhev AG, Hao XJ et al (2011) Modelling and experimental measurements of idealised and light-moderate RCF cracks in rails using an ACFM sensor. NDT&E Int 44:427–437
15. Li P, Zeng LK, Shui AZ et al (2008) Design of forecast system of back propagation neural networks based on MATLAB. Comput Appl Softw 25(4):149–150
16. Li W, Chen GM, Zheng XB (2007) Crack sizing for alternating current field measurement based on GRNN. J Univ Pet China (Edn Nat Sci) 31(2):105–109
17. Cui T, Sun YZ, Xu J et al (2011) Reactive power optimization of power system based on improved niche genetic algorithm. Proc CSEE 31(19):43–50
18. Yang GJ, Cui PY, Li LL (2001) Applying and realizing of genetic algorithm in neural networks control. J Syst Simul 13(5):567–570
19. Li WC, Song DM, Chen B (2006) Artificial neural network based on genetic algorithm. Comput Eng Des 27(2):316–318
20. Li S, Liu LJ, Xie YL (2011) Chaotic prediction for short-term traffic flow of optimized BP neural network based on genetic algorithm. Control and Dec 26(10):1581–1585
21. Li S, Luo Y, Zhang MR (2011) Prediction method for chaotic time series of optimized BP neural network based on genetic algorithm. Comput Eng Appl 47(29):52–55

Identification of Tiny Surface Cracks in a Rugged Weld by Signal Gradient Algorithm Using the ACFM Technique

1 Introduction

The welding procedure is widely used in the manufacturing industry. As a critical connecting portion, it easily introduces cracks in the weld and the HAZ due to the discontinuous material, corrosive environment, varying temperatures, and complex stress. Welds should be inspected routinely using a nondestructive testing (NDT) method in in-service time or before use [1–4].

Magnetic particle testing (MT) and penetrant testing (PT) are effective methods used to detect surface cracks in a weld. However, the surface of the structure should be cleaned thoroughly, including the coating, the attachment, and the greasy dirt [5, 6]. Additionally, the testing result depends on the experience of the operator. Ultrasonic testing (UT) is usually used for the detection of internal defects [7]. Eddy current testing (ET) is easily confused by the lift-off variations, making it hard to identify a tiny crack in a rugged weld [8].

The ACFM technique was developed for the detection of cracks in underwater structures. Due to the advantages of non-contact testing, less cleaning, and quantitative detection, the ACFM technique has been widely used on land and underwater [9–11]. The principle of ACFM is shown in Fig. 1. The excitation coil induces the local uniform current field in the specimen. When a crack is present, the current field is disturbed. The disturbed current field makes the space magnetic field distorted. The magnetic field in the X direction, Bx (parallel to the crack), shows a trough in the center of the crack which contains depth information of the crack. The magnetic field in the Z direction, Bz (perpendicular to the specimen), shows two opposite peaks at the tips of the crack, which reflects the length of the crack. Normally, a so-called butterfly plot (the Bx against the Bz) is presented to decide whether a crack is present or not. In practice, the cracks are identified by the number of the loops in the butterfly plot after detection [12].

However, when the ACFM is used to detect tiny cracks in a weld, the ripples and reinforcements can be regarded as the distance variations between the probe and the

W. Li et al., *Alternating Current Field Measurement Technique for Detection and Measurement of Cracks in Structures*, https://doi.org/10.1007/978-981-97-7255-1_2

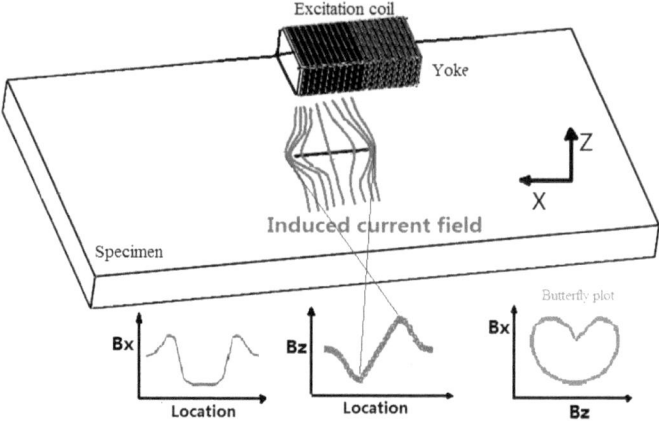

Fig. 1 Principle of ACFM

specimen [13] which causes the lift-off of the probe variations, as shown in Fig. 2. The lift-off variations bring many interference signals, making the signal to noise ratio (SNR) poor [14, 15]. The response signals of a probe are harsh, even on a crack. The *Bx* and *Bz* are confused by the noise caused by the lift-off variations. As a result, the conventional butterfly plot cannot effectively identify tiny cracks. Of course, the size of the crack cannot be evaluated by the *Bx* and *Bz* at this stage. As the first step, how to find a tiny crack in a weld is critical work in the ACFM field.

The probability of detection (POD) using the ACFM technique decreases drastically for tiny cracks whose lengths are less than 10 mm [16]. Smith et al. applied the ACFM technique to inspect the welds in stainless steel nuclear storage tanks [17]. The results showed that most of the tiny cracks (length less than 10 mm, depth less than 1 mm) could not be identified effectively. Mostafavi et al. presented the

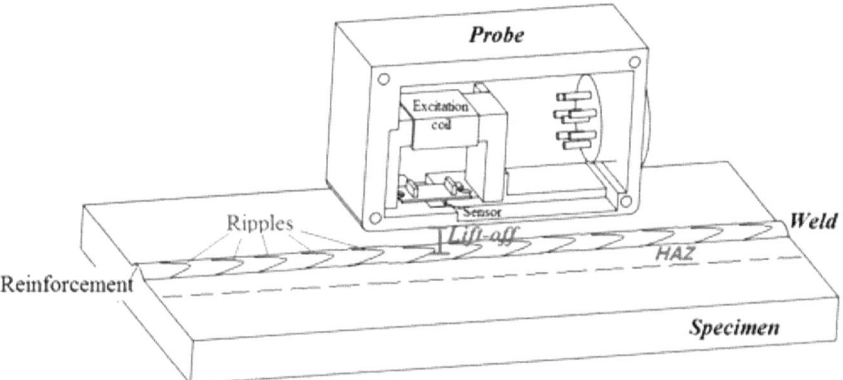

Fig. 2 Weld ripples and reinforcement cause lift-off variations

mathematical method for the theoretical prediction of the magnetic field distribution around a short crack (length 30 mm) [18]. Yuan et al. presented the two-step interpolation algorithm for the measurement of long and short cracks in the pipe string using the uniform alternating current field [19]. Leng et al. proposed the combined metal magnetic memory (MMM) and the ACFM technique for the detection and evaluation of critical weld joints in a jack-up offshore platform [20]. Rowshandel et al. presented the artificial neural network (ANN) for sizing the important subsurface section of the multiple cracks using ACFM [21]. These methods obtained good characteristic signals from the cracks without lift-off variations [22, 23]. However, the characteristic signal and butterfly plot are cluttered when ACFM is used to identify tiny surface cracks in the rugged weld.

The aim of this paper is to find out tiny surface cracks in rugged welds. This paper is organized as follows: in Sect. 2, the insensitive signal to the lift-off variation is pointed out by simulations and experiments; in Sect. 3, the tiny crack is detected in the rugged weld and the signal gradient algorithm is presented to identify the tiny crack; and in Sect. 4, the conclusion and further work are given.

2 Insensitive Signal to Lift-Off Variations

2.1 Simulation Model

In terms of the theory, the Bx is produced by the decrease of the current density in the depth direction of the crack. Thus, the Bx keeps an individual value (background signal) when a crack is not present. The Bz is produced by the deflection of the induced current field, thus the Bz is zero when a crack is not present. When the lift-off varies, the background signal changes—but the current field is still uniform (the current density value changes, but still no deflection). Thus, the Bx is easily affected by the lift-off variations and the Bz is insensitive to lift-off variations.

The 2D simulation model of the ACFM probe was set up as shown in Fig. 3a. In the model, the frequency is 2 kHz and the current amplitude is 50 mA [24]. The material of the yoke is Mn–Zn ferrite and the specimen is a steel plate. The relative permeability of the yoke is 2000. The conductivity and relative permeability of the specimen are 5×10^6 S/m and 200, respectively. Because steel permeability is much greater than air permeability, the magnetic line of force propagates from the U-shaped yoke to the steel. Meanwhile, due to the skin effect, most of the magnetic line of force gathers in the thin surface of the steel. The magnetic field produces a uniform current field in the center of the specimen [25, 26].

When the lift-off increases, more magnetic field leaks into the air. Thus, the background signal of the Bx increases when the lift-off goes up. While there is no magnetic field in the Z direction, the Bz keeps zero because of the uniform current field, as shown in Fig. 3b. We can make a conclusion that the Bx is easily affected by the lift-off variations and the Bz is insensitive to the lift-off variations.

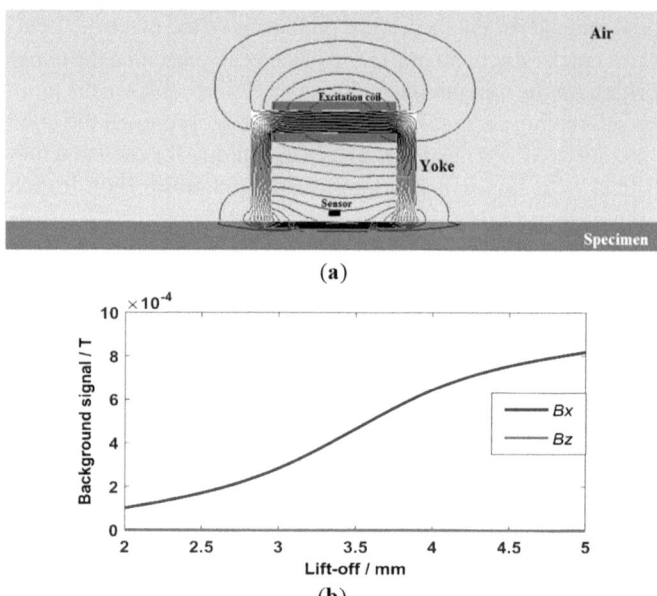

(a)

(b)

Fig. 3 A 2D simulation model of ACFM probe. **a** Distribution of magnetic field; **b** background signal with different lift-off

2.2 System Set Up

The ACFM testing system is shown in Fig. 4a. The system consists of three main parts: a probe, a signal processing box, and a 3-axis scanner. The probe includes an excitation coil, a MnZn-ferrite yoke, a detection sensor, and a primary processing circuit, as shown in Fig. 4b. The excitation coil is wound on the MnZn-ferrite yoke with 500 turn enameled wires whose diameter is 0.15 mm. The detection sensor is a 2-axis tunnel magneto resistance (TMR) sensor that is set at the bottom center of the yoke [27, 28]. The 2-axis TMR sensor (Type: TMR2303, Capacity: $\pm$ 80 Oe, Sensitivity: 3 mV/Oe, made by MULTI DIMENSION, China), is used to measure the Bx and Bz at the same time. The Bx and Bz are amplified 50 times and 100 times, respectively, by the primary processing circuit. The thickness of the probe shell bottom is 1 mm to keep the lift-off of the sensor 1 mm. The probe is installed on a 3-axis scanner which is controlled by the programmable logic controller (PLC).

The signal processing box includes a power module, an excitation module, a filtration module, and an acquisition module, as shown in Fig. 4b. The power module provides the power source to keep the modules running for more than six hours. The excitation module produces a sinusoidal signal, whose frequency is 2 kHz and amplitude is 10 V. The sinusoidal signal is loaded on the excitation coil in the probe. The Bx and Bz are filtered by the filtration module using a low-pass filter whose cut-off frequency is 20 kHz. The Bx and Bz are gathered by the acquisition module and

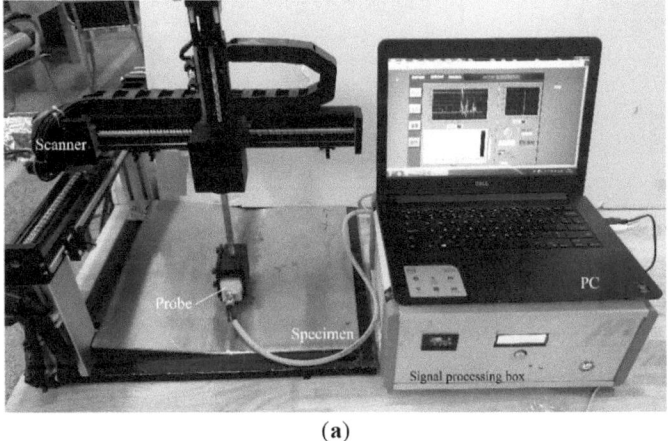

(a)

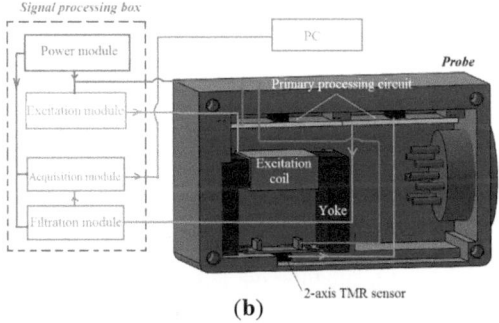

(b)

Fig. 4 ACFM system. **a** Photo of the system; **b** probe and signal processing module

transmitted to the personal computer (PC) for the signal processing and analyzing software.

2.3 Lift-Off Variation Experiments

To verify the insensitive signal to lift-off variations, the probe is driven up and down above the specimen by the scanner. The first specimen is a mild steel plate with a semi-elliptical artificial crack (length: 50 mm, width: 0.5 mm and depth: 5 mm). There is no weld on the plate and the surface is flat. In the lift-off upward testing experiments, the probe is driven by the scanner to scan along the artificial crack with a lift-off of 1 mm at a speed of 2 mm/s. The probe is raised up 1 mm by the scanner and then dropped after 2 s, as shown in Fig. 5a.

As shown in Fig. 5b, the **Bx** shows a trough in the center of the crack, while the **Bz** shows opposite peaks at the tips of the crack. The characteristic signals are

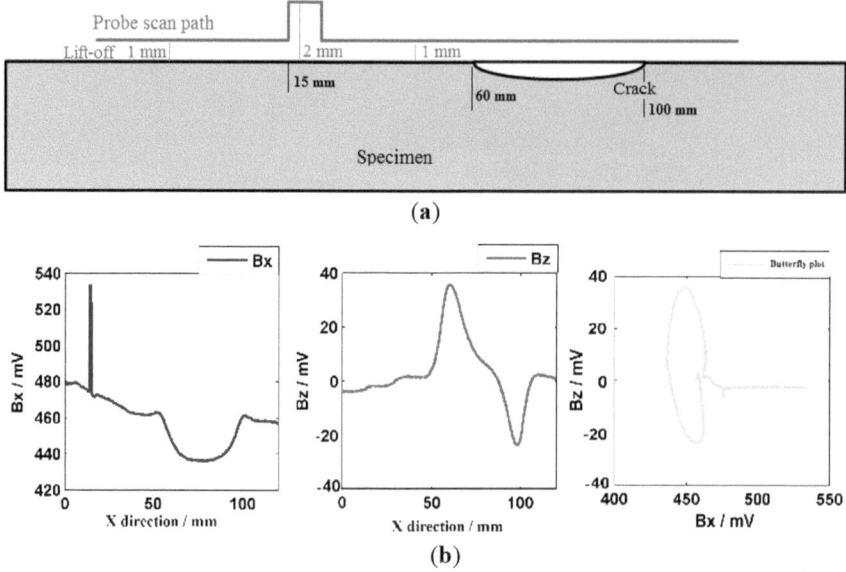

Fig. 5 Testing results when lift-off upward. **a** Probe scan path; **b** *Bx*, *Bz*, and butterfly plot

consistent with ACFM theory. When the lift-off is upward, the **Bx** shows a maximum peak and the **Bz** remains the same. This is because the background signal increases when the lift-off goes up. Meanwhile, the induced current field is still uniform and the **Bz** almost remains the same. There are obvious loops in the butterfly plot for the Identification of a crack in the flat specimen. The **Bx** distorted value caused by the 1 mm lift-off variation is much larger than that caused by the crack, which produces an interference signal in the butterfly plot.

In the lift-off downward testing experiments, the probe was lowered down 1 mm by the scanner and then raised after 2 s, as shown in Fig. 6a. When the lift-off is downward, the **Bx** shows a minimum trough and the **Bz** almost remains the same, as shown in Fig. 6b. This is because the background signal decreases when the lift-off goes down. In the same way, the crack can be identified by the butterfly plot. The lift-off downward also produces an obvious interference signal in the butterfly plot.

To simulate the uneven surface of a weld, plastic humps were set on the surface of the specimen in front of the crack, as shown in Fig. 7a. When the probe scans the humps, the lift-off varies seriously from 1 to 4 mm. The characteristic signals are shown in Fig. 7b. The **Bx** shows hash signals, while the **Bz** varies slightly in the hump area. The results show that the **Bx** is affected seriously by the lift-off variations and the **Bz** is insensitive to the lift-off variations. The **Bz** can be set as the insensitive signal to identify the tiny surface crack in the weld.

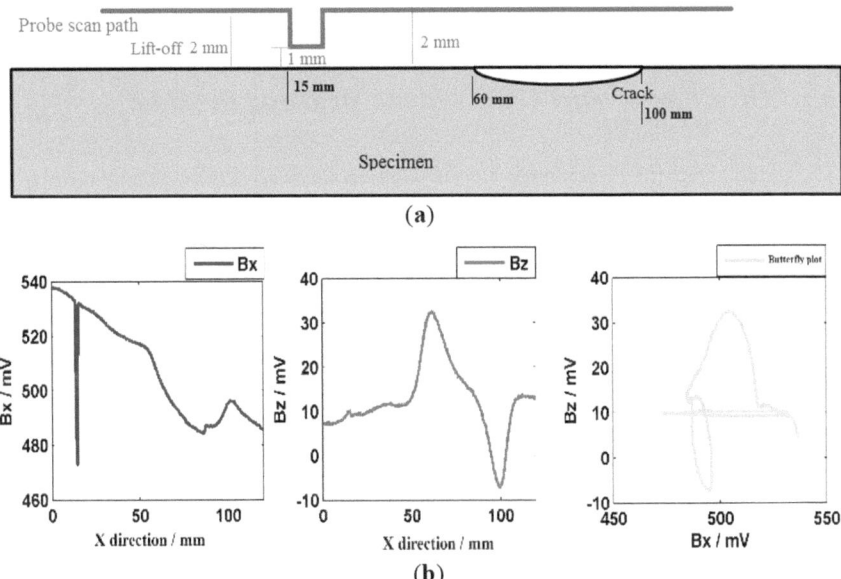

Fig. 6 Testing results when probe downward. **a** Probe scan path; **b** *Bx*, *Bz*, and butterfly plot

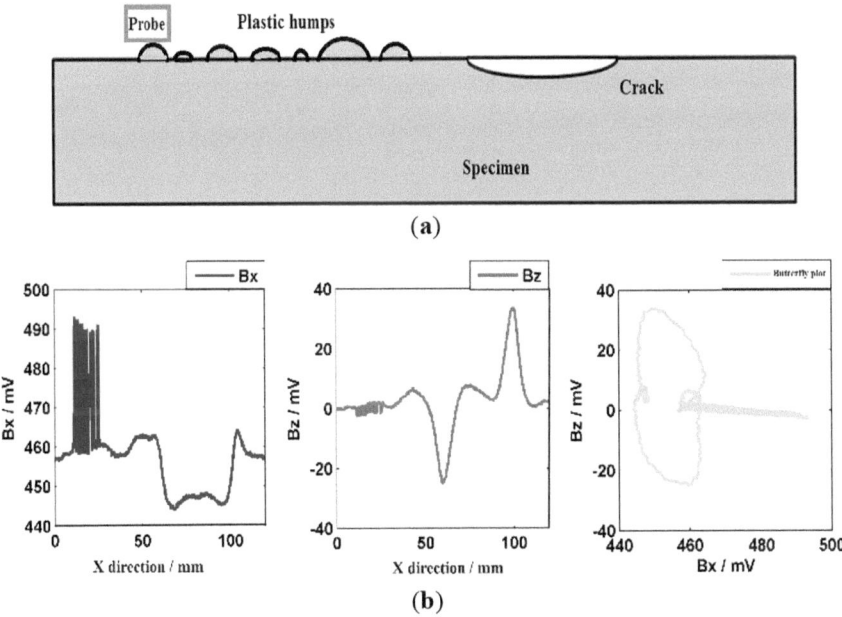

Fig. 7 Testing results when lift-off varies seriously. **a** Probe scan path; **b** *Bx*, *Bz*, and butterfly plot

3 Signal Gradient Algorithm

3.1 Detection of Tiny Cracks in the Weld and the HAZ

The second specimen is a plate with one weld, as shown in Fig. 8. The material of the specimen is the same as the first specimen. The thickness of the plate is 4 mm. The width of the weld is 10 mm and the height of the weld reinforcement is 2 mm. (The distance between the weld and the HAZ is 2 mm.) The weld ripples are rugged and the thickness of the ripples range from 0.1 to 1 mm. Thus, when the ACFM probe scans along the weld, the lift-off of the probe varies from 0.1 to 1 mm.

There are three tiny surface cracks in the weld with the same length (4 mm) and different depths (3.5, 3.0 and 2.5 mm). There are other 3 cracks with identical length of 4 mm and the same depths (W1, W2, and W3) in the HAZ. The depths of the cracks are given in Table 1.

The probe is driven by the scanner to scan the weld from W1 to H3 at a speed of 5 mm/s. The probe can shake up and down with the ripples. The testing results are shown in Fig. 9a. The *Bx* is chaotic because there are many interference signals caused by the lift-off variations. Thus, the cracks cannot be identified clearly by the *Bx* signal. There are six characteristic signals in the *Bz* when a crack is present. The SNR of the *Bz* is much better than that of the *Bx,* because the *Bz* is insensitive to the lift-off variations. As shown in Fig. 9b, the butterfly plot is irregular, which cannot be used to identify the 6 tiny cracks.

The third specimen is shown in Fig. 10. There are two different length cracks in the HAZ and three different length cracks in the weld. The cracks are at the same depth (3.0 mm). The lengths of the cracks are shown in Table 2.

The weld was scanned at the same speed, and the testing results are shown in Fig. 11. The *Bx* shows two troughs, and the *Bz* shows two troughs and peaks at

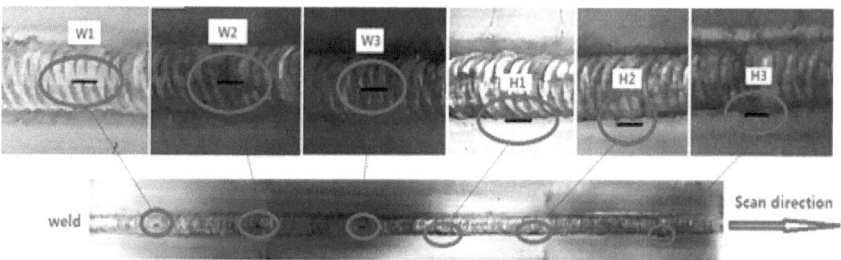

Fig. 8 The second specimen with different depth cracks

Table 1 The depth of the cracks

No.	W1	W2	W3	H1	H2	H3
Depth/mm	3.5	3.0	2.5	3.5	3.0	2.5

Fig. 9 Characteristic signals of different depth cracks in the weld and the HAZ. **a** Bx and Bz; **b** butterfly plot

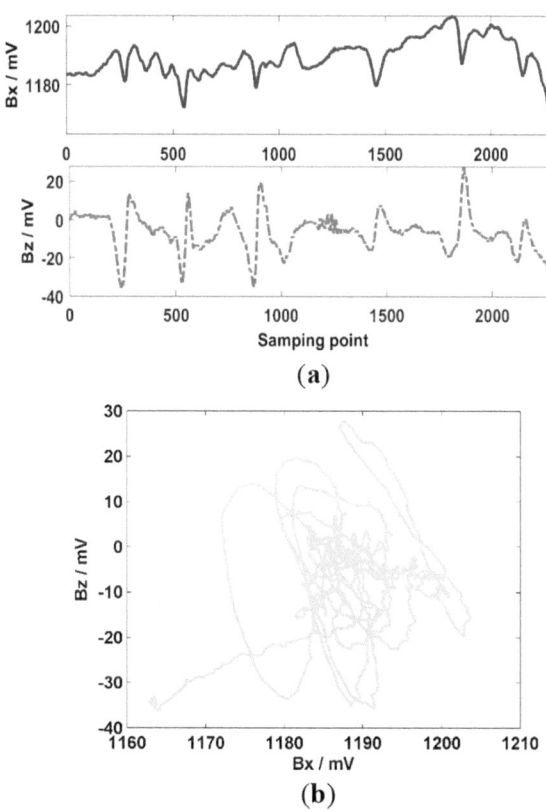

(a)

(b)

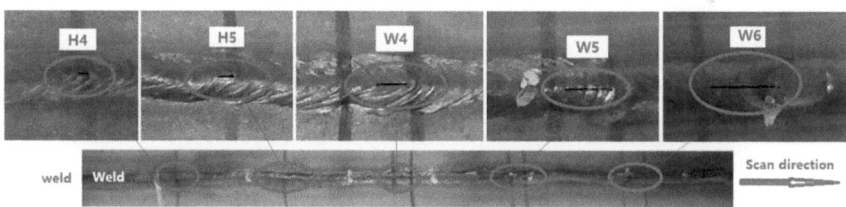

Fig. 10 The third specimen with different length cracks

Table 2 The length of the cracks

No.	H4	H5	W4	W5	W6
Length/mm	1.0	2.0	4.0	6.0	8.0

Fig. 11 Characteristic
signals of different length
cracks in the weld and the
HAZ. **a** Bx and Bz;
b butterfly plot

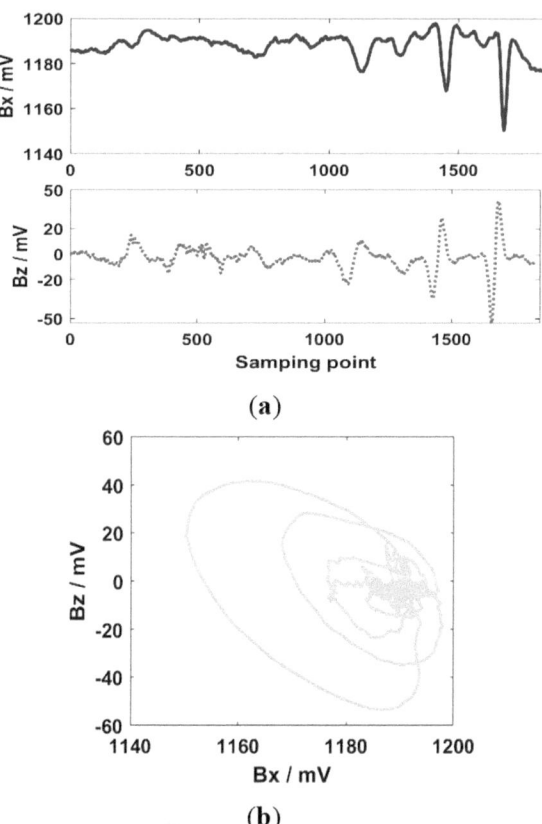

(a)

(b)

the W5 and W6 cracks. The characteristics of the **Bx** and **Bz** are incomplete at the
W4 crack. For the H4 and H5 cracks, there is one response in the **Bx**, while the **Bz**
shows slight perturbations because of the very short length of the crack. As shown
in Fig. 11b, there are two obvious loops in the butterfly plot, which can be used to
identify the last two longer cracks in the weld. Other cracks cannot be identified
effectively by the irregular tracks in the butterfly plot.

3.2 Signal Gradient Algorithm

The gradient method can find the amplitude change rate of the signal. The **Bz** changes
are small when lift-off varies. Thus, the gradient of the **Bz** will be zero when a crack is
not present. What's more, there are opposite peaks at the two ends of the crack in the
Bz. As a result, a slight perturbation in the **Bz** caused by the crack is amplified many
times by the gradient method. Thus, the gradient of the **Bz** will show an outstanding

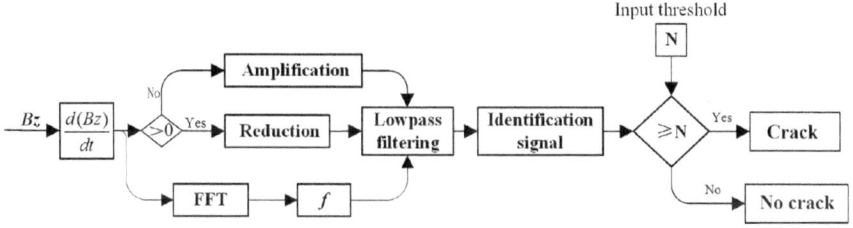

Fig. 12 Signal gradient algorithm for the processes of the Bz

peak in the center of the crack. The SNR of the crack identification signal will be improved greatly in the near zero background signal.

The signal gradient algorithm is presented to process the Bz by the following steps, as shown in Fig. 12. First, the gradient of the Bz is calculated in real time. If a crack is not present, the value is near zero. When a crack is present, the gradient of the Bz will show some negative values and a major positive value. Then the negative values are decreased and the positive value is magnified. To make the curve smooth, the signal is processed by a 6-th order Butterworth low-pass filter. Thus, the identification signal is presented with a great positive peak and near zero value. The threshold is set beforehand to compare the identification signal with the threshold automatically. If the identification signal is greater than the threshold, the crack can be identified effectively because of the high SNR.

The Bz signals of different depth cracks (from Fig. 9a) and different length cracks (from Fig. 11a) are processed by the signal gradient algorithm. The identification signals of the tiny surface cracks with lift-off variations in the weld and the HAZ are shown in Fig. 13. As shown in Fig. 13a, there are six obvious peaks in the identification signals caused by the different length cracks. Meanwhile, there are three obvious peaks caused by the three longer cracks in the weld and two slight peaks caused by the two shorter cracks in the HAZ, as shown in Fig. 13b. To remove the background signal and identify the tiny cracks effectively in real time, the threshold can be set in advance using 1.5 times the maximum noise amplitude in the identification signal. In these experiments, the maximum noise amplitude was 53 and the threshold was set as 80. Thus, all six different depth cracks and five different length cracks can be identified effectively, as shown in Fig. 13c, d, respectively.

3.3 Discussions

The SNR of the Bz signal is better than that of the Bx in the crack testing results, as shown in Fig. 9a. This is because the Bx signals are greatly disturbed by the lift-off variations caused by the rugged weld ripples and reinforcements. The Bz is relatively stable because it is insensitive to the lift-off variations—thus, the Bz is used as the characteristic signal to identify the tiny crack with lift-off variations. As shown in

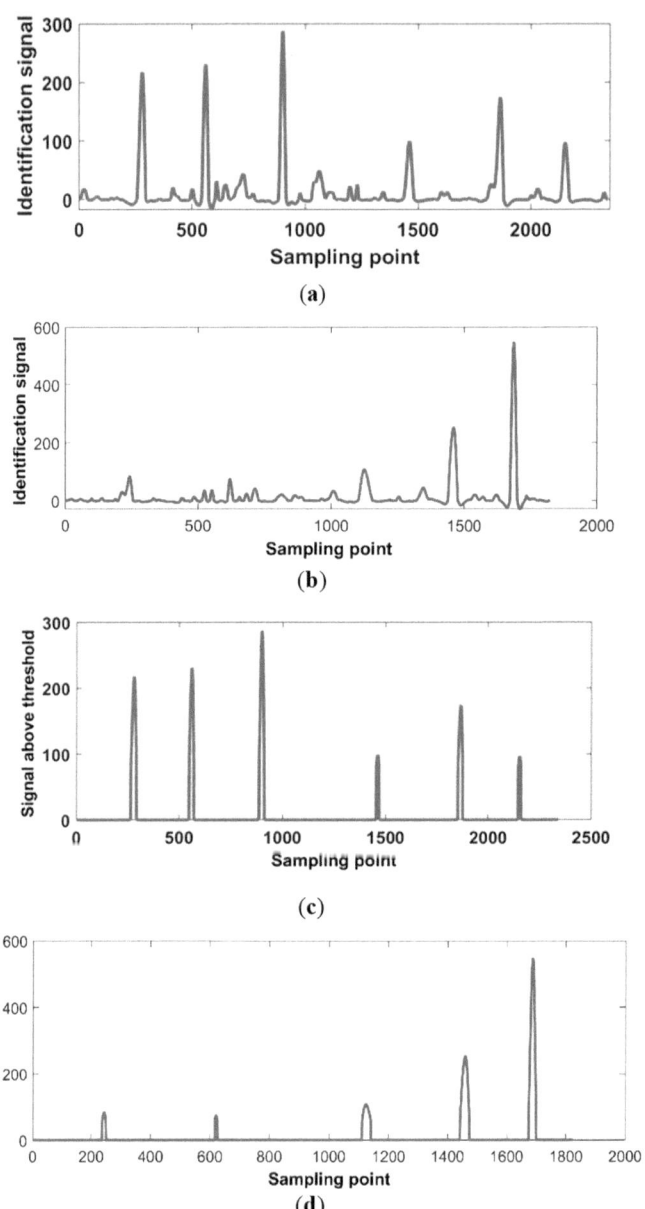

Fig. 13 Identification of tiny cracks by signal gradient algorithm. **a** Identification signal of different depth cracks; **b** identification signal of different length cracks; **c** identification of different depth tiny cracks; **d** identification of different length tiny cracks

Fig. 11a, where the length of the crack is shorter than 4 mm, the distortion of the Bx and Bz decays seriously. This is because the disturbance of the induced current field declines sharply around the short cracks. For the 1 and 2 mm length cracks, the induced current field turns slightly and the magnetic field distorts weakly.

As shown in Fig. 13a, the last three peaks are less than the first three peaks. This is because the last three cracks are in the HAZ and the distance between the probe and the HAZ is greater than the distance between the probe and the weld. Because the first three cracks are in the rugged weld, the background signal in the identification signal has a bit more clutter than that of the last three cracks. As shown in Fig. 13b, the peaks of the three longer cracks are obvious and the peaks of first two shorter cracks are weaker. Because the length of the tiny cracks has a significant impact on the Bx and Bz signals, it is difficult to identify the shorter cracks (less than 2 mm) in the weld and the HAZ—the shorter cracks are easily covered by the noise signal. Overall, in these experiments the SNR of the identification signals must be relatively high to identify all the tiny cracks in the weld and the HAZ by threshold compared to the conventional butterfly plot.

4 Conclusions

In this paper, a signal gradient algorithm was presented to identify tiny cracks in welds using the ACFM technique. The insensitive signal to the lift-off variations was pointed out by the simulations and experiments. The ACFM probe with a 2-axis TMR sensor and testing system were developed. The different depth and different length tiny surface cracks were detected in the rugged weld and the HAZ using an ACFM testing system. The results show that the Bz signal was the insensitive signal to the lift-off variations. The SNR of the crack response signal was greatly improved by the signal gradient algorithm. Thus, all the tiny surface cracks can be effectively identified in the weld and the HAZ by the signal gradient algorithm using the ACFM technique. Further work will focus on the identification of other type defects, evaluation of tiny defects in the weld, and the HAZ.

References

1. Taheri H, Kilpatrick M, Norvalls M, Harper WJ (2019) Investigation of nondestructive testing methods for friction stir welding. Metals 9:624
2. Wang R, Liu ZH, Wu JF, Jiang BY, Li B (2019) Research on phased array ultrasonic testing on CFETR vacuum vessel welding. Fusion Eng Des 139:124–127
3. Gao XD, Du LL, Ma N, Zhou XH, Wang CY, Gao PP (2019) Magneto-optical imaging characteristics of weld defects under alternating and rotating magnetic field excitation. Opt Laser Technol 112:188–197

4. Andreaus U, Baragatti P, Casini P, Iacoviello D (2016) Experimental damage evaluation of open and fatigue cracks of multi-cracked beams by using wavelet transform of static response via image analysis. Struct Control Health Monit 105:227–244
5. Shelikhov GS, Glazkov YA (2011) On the improvement of examination questions during the nondestructive testing of magnetic powder. Russ J Nondestr Test 47:112–117
6. Pashagin AI, Shcherbinin VE (2012) Indication of magnetic fields with the use of galvanic currents in magnetic-powder nondestructive testing. Russ J Nondestr Test 48:528–531
7. Takiy AE, Kitano C, Higuti RT, Granja SCG, Prado VT, Elvira L, Martinez-Graullera Ó (2017) Ultrasound imaging of immersed plates using high-order lamb modes at their low attenuation frequency bands. Mech Syst Sign Pr 96:321–332
8. Ricci M, Silipigni G, Ferrigno L, Laracca M, Adewale I, Tian GY (2017) Evaluation of the lift-off robustness of eddy current imaging techniques. NDT&E Int 85:43–52
9. Katoozian D, Hasanzadeh RPR (2017) A fuzzy error characterization approach for crack depth profile estimation in metallic structures through ACFM data. IEEE T Magn 53:6202100
10. Lewis AM, Michael DH, Lugg MC, Collins R (1988) Thin-skin electromagnetic fields around surface-breaking cracks in metals. J Appl Phys 64:3777–3784
11. Raine A (2002) Cost benefit applications using the ACFM technique. Insight 44:180–181
12. Akbari-Khezri A, Sadeghi SHH, Moini R, Sharifi M (2016) An efficient modeling technique for analysis of AC field measurement probe output signals to improve crack detection and sizing in cylindrical metallic structures. J Nondestr Eval 35:9
13. Deng ZY, Sun YH, Yang Y, Kang YH (2017) Effects of surface roughness on magnetic flux leakage testing of micro-cracks. Meas Sci Technol 28:045003
14. Lu MY, Xu HY, Zhu WQ, Yin LY, Qian Z, Peyton AJ, Yin WL (2018) Conductivity lift-off invariance and measurement of permeability for ferrite metallic plates. NDT&E Int 95:36–44
15. Andreaus U, Casini P, Vestroni F (2005) Nonlinear features in the dynamic response of a cracked beam under harmonic forcing. In: DETC'05, proceeding of the ASME 2005 international design engineering technical conferences & computers and information in engineering conference, Long Beach, CA, USA, 24–28 Sept 2005, pp 2083–2089
16. Shen JL, Zhou L, Rowshandel H, Nicholson GL et al (2015) Determining the propagation angle for non-vertical surface-breaking cracks and its effect on crack sizing using an ACFM sensor. Meas Sci Technol 26:115604
17. Smith M, Laenen C (2007) Inspection of nuclear storage tanks using remotely deployed ACFMT. Insight 49:17–20
18. Mostafavi RF, Mirshekar-Syahkal D (1999) AC fields around short cracks in metals induced by rectangular coils. IEEE T Magn 35:2001–2006
19. Yuan XA, Li W, Chen GM, Yin XK, Yang WC, Ge JH (2018) Two-step interpolation algorithm for measurement of longitudinal cracks on pipe strings using circumferential current field testing system. IEEE T Ind Inform 14:394–402
20. Leng JC, Tian HX, Zhou GQ, Wu ZM (2017) Joint detection of MMM and ACFM on critical parts of jack-up offshore platform. Ocean Eng 35:34–38 (in Chinese)
21. Rowshandel H, Nicholson GL, Shen JL, Davis CL (2018) Characterization of clustered cracks using an ACFM sensor and application of an artificial neural network. NDT&E Int 98:80–88
22. Yuan XA, Li W, Chen GM, Yin XK, Jiang WY, Zhao JM, Ge JH (2019) Inspection of both inner and outer cracks in aluminum tubes using double frequency circumferential current field testing method. Mech Syst Sign Pr 127:16–34
23. Tian GY, Sophian A (2005) Reduction of lift-off effects for pulsed eddy current. NDT&E Int 38:319–324
24. Yuan XA, Li W, Chen GM, Yin XK, Ge JH (2017) Frequency optimisation of circumferential current field testing system for highly-sensitive detection of longitudinal cracks on a pipe string. Insight 59:378–382
25. Li Y, Yan B, Li D, Li YL, Zhou DQ (2016) Gradient-field pulsed eddy current probes for imaging of hidden corrosion in conductive structures. Sens Actuators A Phys 38:251–265
26. Li Y, Ren ST, Yan B, Abidin IMZ, Wang Y (2017) Imaging of subsurface corrosion using gradient-field pulsed eddy current probes with uniform field excitation. Sensors 17:1747

27. Atzlesberger J, Zagar BG, Cihal R, Brummayer M, Reisinger P (2013) Sub-surface defect detection in a steel sheet. Meas Sci Technol 24:084003
28. AbdAlla AN, Faraj MA, Samsuri F, Rifai D, Ali K, Al-Douri Yarub (2019) Challenges in improving the performance of eddy current testing. Rev Meas Control 52:46–64

Visual Evaluation of Irregular Cracks in Steel by Double Gradient Fusion Algorithm Using Composite ACFM-MFL Testing Method

1 Introduction

The steel is the most common material in the industrial and civil facilities. Because of the corrosion, stress and fatigue, the cracks are easily introduced in the surface of the steel. Generally, the incipient crack grows and tends to different directions resulting in the irregular crack due to the complex stress and special structural shape [1–3]. In order to prevent the corrosion, there are various coatings on the surface of the steel. The irregular crack is covered by the coating, which leaves hidden trouble in the facility [4]. Thus it is of prime significance to propose an effective nondestructive testing (NDT) method to achieve visual detection and evaluation of the irregular crack under the coating [5, 6].

Usually, several NDT methods are proposed to detect and evaluate the crack in the steel. The visual testing (VT), magnetic particle testing (MT) and penetrant testing (PT) are common methods to inspect the surface crack [7]. The surface morphology of the crack can be shown visually by the gathered magnetic particle, penetrant or optical instrument [8, 9]. However, the attachments, contaminations and coatings should be cleaned thoroughly. After inspection, the coating should be painted again, which is time-consuming and high-cost. What's more, these NDT methods cannot be applied in the service time in most case. The ultrasonic testing (UT) is mainly used to test the inner defects in the steel, which needs coupling medium during application [10]. Because of the radioactivity, the radiographic testing (RT) is forbidden in many industrial field.

The electromagnetic nondestructive testing (ENDT) techniques are excellent methods to test the surface crack due to the advantages of high sensitivity, non-contact, low cost and no coupling required [11–14]. Many scholars have made broad scale researches on the detection and evaluation of cracks using the ENDT technique. Rowshandel, et al. proposed the artificial neural network (ANN) to learn the inverse relationship between the crack pocket length and the ACFM signal for a given cluster of rolling contact fatigue (RCF) cracks in the railway rails [15, 16].

© The Author(s) 2025

W. Li et al., *Alternating Current Field Measurement Technique for Detection and Measurement of Cracks in Structures*, https://doi.org/10.1007/978-981-97-7255-1_3

Many scholars have made broad scale researches on the inspection of cracks using the ACFM technique. Dover, Lewis, et al. developed the classical theoretical model of the ACFM technique [17, 18]. The model is developed on the assumption that the injected current field flows around one crack vertically. Pasadas, et al. presented the Tikhonov regularization inversion algorithm to obtain the geometrical profile of defects in the 2D surface view [19]. Feng, et al. proposed an optimized ACFM probe for the detection of inner cracks inside the pipeline [20].

These works have obtained good results of surface cracks when the induced current field is perpendicular to the crack. In other words, the probe should scan along the direction of the crack. However, the direction of the crack is unclear under the coating before the inspection in practical work. Thus, some scholars did research on the inspection of different angle cracks. Rowshandel, et al. proposed the artificial neural network (ANN) to learn the inverse relationship between the crack pocket length and the ACFM signal for a given cluster of rolling contact fatigue (RCF) cracks in the railway rail [21]. Shen, et al. studied the relationship between ACFM signals and propagation angles of surface-breaking cracks [22]. In our previous work, the rotating alternating current field measurement (RACFM) technique was presented to overcome the limitations of the classical theoretical model of the ACFM technique. The rotating current field is induced by two orthogonal excitation coils [23]. Thus, the induced current field is always perpendicular to arbitrary-angle cracks. These works all focus on the inspection of one crack with some specific angles [24, 25]. However, the irregular crack usually grows and tends to different directions, which results in some partial cracks. The development of the irregular crack should be monitored by the periodic detection. On one hand, the rotating current field cannot pass through the irregular vertically always because the rotating current field is cut off by the discontinuous area. On the other hand, the RACFM inspects cracks by the conventional amplitude of the magnetic signal, which cannot show the surface morphology of the crack visually [26]. What's more, two excitation coils and excitation signals should be used in the RACFM system making the RACFM probe huge and complicated.

In this paper, the double gradient fusion algorithm is presented to achieve visual evaluation of the irregular crack under the coating by the composite ACFM-MFL testing method. On one hand, the excitation coil can excite AC primary magnetic field to produce the leaked magnetic field caused by the transverse crack. The primary magnetic field induces current field in the steel to produce the distorted magnetic field caused by the longitudinal crack. Thus the magnetic field can be measured by the same probe with AC excitation, which is beneficial to achieve visual evaluation of the irregular crack. On the other hand, the composite ACFM-MFL testing method is sensitive to the orthogonal directional cracks. As a result, any part of the irregular crack can be detected and evaluated effectively.

The rest of the paper is organized as follows. In Sect. 2, the FEM model of the ACFM-MFL with an irregular crack is presented to analyze the leaked and distorted magnetic field. In Sect. 3, the double gradient fusion algorithm is developed to reconstruct the morphology of the irregular crack using the simulation results. In Sect. 4, the irregular crack testing experiments are carried out and the irregular crack is evaluated visually by the double gradient fusion algorithm. Finally, the conclusion and future work are drawn in Sect. 5.

2 Finite Element Method Model

The FEM model of the ACFM-MFL with the irregular crack is set up, as shown in Fig. 1. The model includes an excitation coil, a U-shaped yoke, a steel specimen and an irregular crack. The excitation coil is wound on the U-shaped yoke with 500 turns. The lift-off of the U-shaped yoke is 3 mm to the specimen (1 mm is for the shell of the probe and the lift-off of the sensor is 2 mm). The irregular crack is consists of 4 cracks whose angle is 0°, 30°, 60°, 90°. The depth and length of the 4 cracks are 6 mm and 30 mm respectively. The material parameters of the model is given in Table 1.

The excitation coil is loaded by a sine signal whose frequency is 1000 Hz and current magnitude is 50 mA. According to the Lenz's law, the primary magnetic field is excited in the **X** direction, while the induced uniform current field flows in the

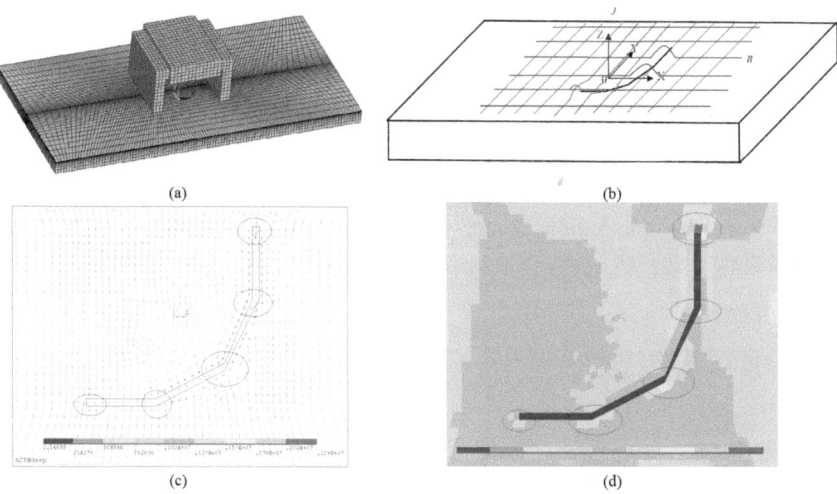

(a)

(b)

(c)

(d)

Fig. 1 FEM model of ACFM-MFL with irregular crack. **a** Model FEM. **b** Principle of distorted magnetic field around irregular crack. **c** Vector current field around irregular crack. **d** Current density around irregular crack

Table 1 Parameters of EFM model

Model	Air	Excitation coil	Steel specimen
Resistivity/Ω m	0.5×10^4	1.7×10^{-8}	9.78×10^{-8}
Relative permeability/μ_r	1.0	1.0	1000

Y direction, as shown in Fig. 1b. When the irregular crack is present, the primary magnetic field pass through the 90° crack, 60° crack and 30° crack. As a result, the primary magnetic field leaks in the air and produces the magnetic leakage signal. Meanwhile, the induced uniform current field is disturbed by the irregular crack, as shown in Fig. 1c, d. Especially, the current is interrupted by the 0° crack, 30° crack and 60° crack, which makes the space magnetic field distorted. For the 30° and 60° cracks, both the magnetic leakage effect and current field perturbation effect exist at the same time [27].

The space magnetic field is extracted above the irregular crack with the lift-off of 2 mm, as shown in Fig. 2. The magnetic field in the **X** direction (**Bx**) shows the maximum peak at the 90° crack due to the strongest magnetic leakage effect. Because the magnetic leakage effect is much greater than the current field perturbation effect, the trough of **Bx** caused by the decreased current density (Secondary magnetic field) is very tiny relatively at the 0° crack, as shown in Fig. 2a. The magnetic field in the **Z** direction (**Bz**) shows peak and trough at each side of the 90° crack, as shown in Fig. 2b. This is due to the opposite direction of the leaked magnetic field in the **Z** direction [28]. Meanwhile, the surface current field turns around the irregular crack greatly and gathers at the tips of the irregular crack. Thus, the disturbed current field makes the space magnetic field distorted along the 0° crack, 30° crack and 60° crack, as shown in Fig. 2a.

On one hand, because the **Bz** is mainly caused by the surface disturbed current around the irregular crack and the leaked magnetic field besides the crack, the outline of the **Bz** image is more similar to the morphology of the irregular crack (Comparing Fig. 2b with Fig. 2a). On the other hand, the **Bz** is insensitive to the lift-off effect as the zero background signal, which has a higher signal to noise ratio (SNR). Thus the **Bz** is more beneficial to detect and evaluate the irregular crack under the coating. The distorted magnetic field is measured by the magnetic sensor no matter what's the magnetic leakage effect or current field perturbation effect. Thus the **Bz** signal

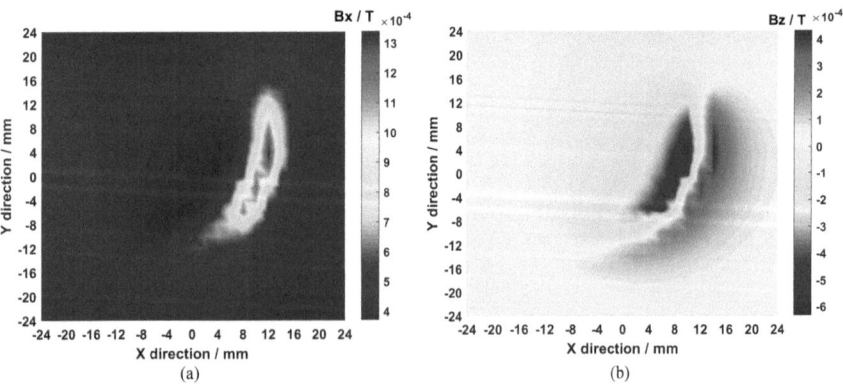

Fig. 2 Distorted magnetic field around irregular crack. **a** Magnetic field in X direction. **b** Magnetic field in Z direction

can be used to reconstruct the morphology of the irregular crack using the composite ACFM-MFL testing method.

3 Double Gradient Fusion Algorithm

As the **Bz** signal changed greatly from the trough to the peak at each side of the irregular crack, the gradient field can be used to find the maximum variation direction and value of the scalar field, which is defined in Eq. (1) [29–32].

$$\boldsymbol{grad}\, u(x, y, z) = \left(\frac{\partial u}{\partial x}, \frac{\partial u}{\partial y}, \frac{\partial u}{\partial z} \right) \tag{1}$$

where the $u(x, y, z)$ is the scalar field, the $\boldsymbol{grad}\, u(x, y, z)$ is the gradient field of the scalar field.

Because there are 4 cracks from 0° to 90°, the peaks and troughs change greatly in the **X** direction and **Y** direction respectively. Thus the **X** direction gradient field $(\mathbf{GX_{Bz}})$ and **Y** direction gradient field $(\mathbf{GX_{Bz}})$ of the **Bz** scalar field (**Bz** value at each location from Fig. 2b) should be calculated using the Eq. (1), as shown in Fig. 3. The results show that the 90° crack, 60° crack and 30° crack can be presented obviously in the **X** direction gradient of the **Bz** scalar field, as shown in Fig. 3a. Meanwhile the 0° crack, 30° crack and 60° crack can be presented in the **Y** direction gradient of the **Bz** scalar field, as shown in Fig. 3b.

To get the whole morphology of the irregular crack, the back ground signals of the $\mathbf{GX_{Bz}}$ and $\mathbf{GY_{Bz}}$ are removed to form the $\mathbf{GX_{RB}}$ and $\mathbf{GY_{RB}}$ respectively. Then $\mathbf{GX_{RB}}$ and $\mathbf{GY_{RB}}$ are normalized to 0–1 to form the $\mathbf{GX_{0-1}}$ and $\mathbf{GY_{0-1}}$ respectively. The maximum distorted value is set as 1 and the minimum distorted value is set as 0, as shown in Fig. 4.

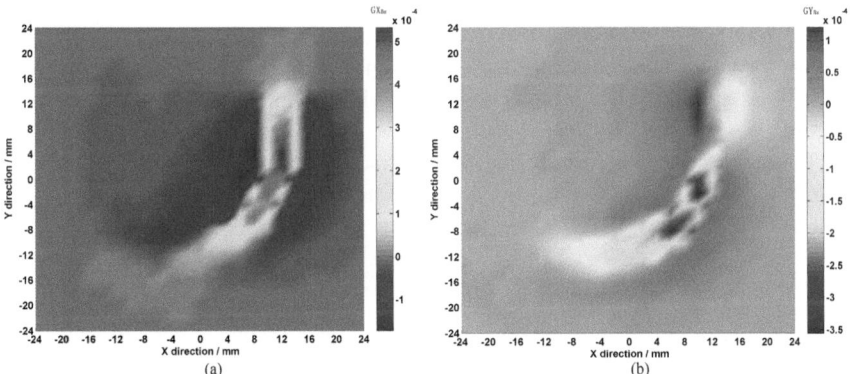

Fig. 3 Double direction gradient field of **Bz**. **a** X direction gradient field. **b** Y direction gradient field

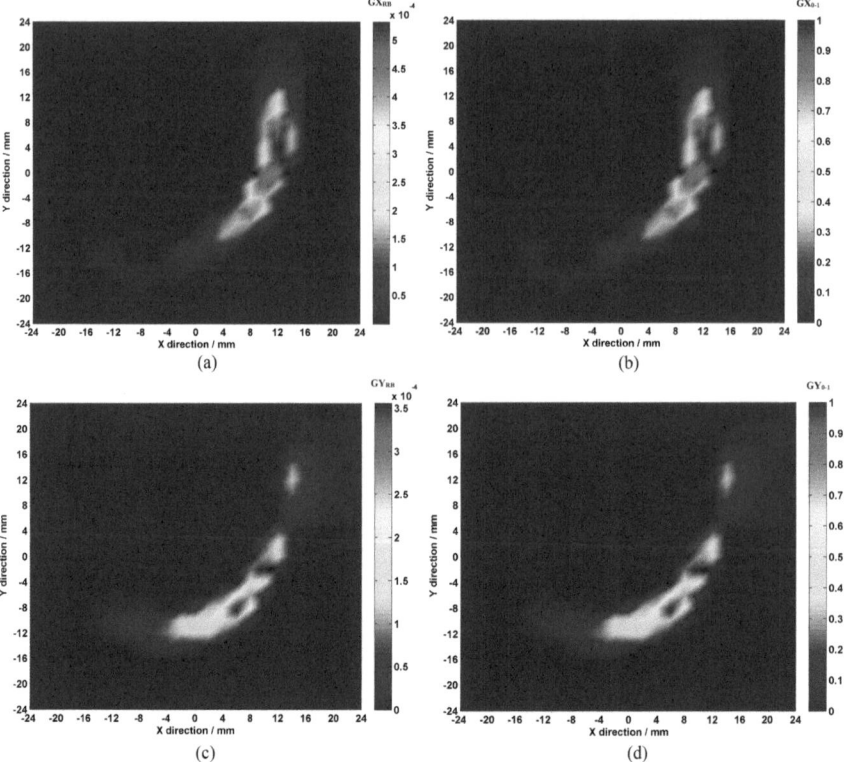

Fig. 4 Normalized gradient field. **a** Removed background signal of **X** direction gradient field. **b** Normalized **X** direction gradient field. **c** Removed background signal of **X** direction gradient field. **d** Normalized **Y** direction gradient field

Then the $\mathbf{GX_{0\text{-}1}}$ and $\mathbf{GY_{0\text{-}1}}$ are added up, as shown in Fig. 5a. The outline of the added gradient is similar to the morphology of the irregular crack. In the end, the added gradient image is converted to the greyscale map. Thus the irregular crack can be shown visually and obviously, as shown in Fig. 5b. We can make a conclusion that the irregular crack can be reconstructed visually by the double gradient fusion algorithm.

Based on the analysis above, the double gradient fusion algorithm is presented in Fig. 6. The double gradient fusion algorithm includes 6 steps:

- Step 1. Get the scalar field $\mathbf{Bz} = \begin{bmatrix} a_{11}\, a_{12} \cdots a_{1n} \\ a_{21}\, a_{22} \cdots a_{2n} \\ a_{31}\, a_{32} \cdots a_{3n} \\ \cdots \cdots \cdots \\ a_{m1}\, a_{m2} \cdots a_{mn} \end{bmatrix}$. a_{mn} is the value of the $\mathbf{Bz}$

  at each location above the irregular crack.

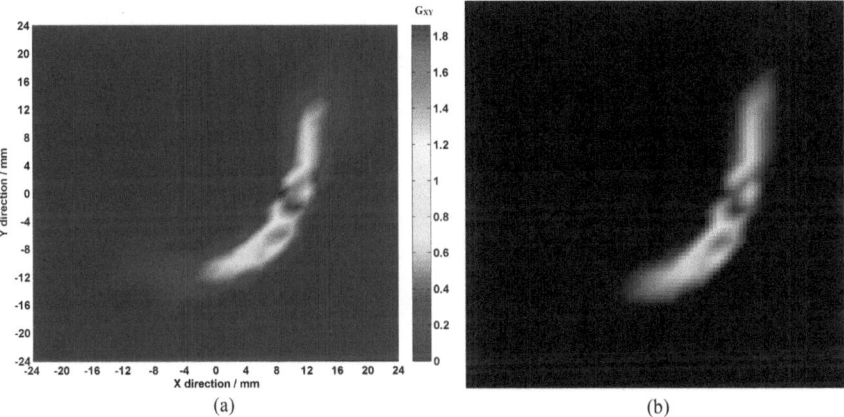

(a)   (b)

Fig. 5 Visual results of irregular crack. **a** Added gradient field. **b** Greyscale map of irregular crack

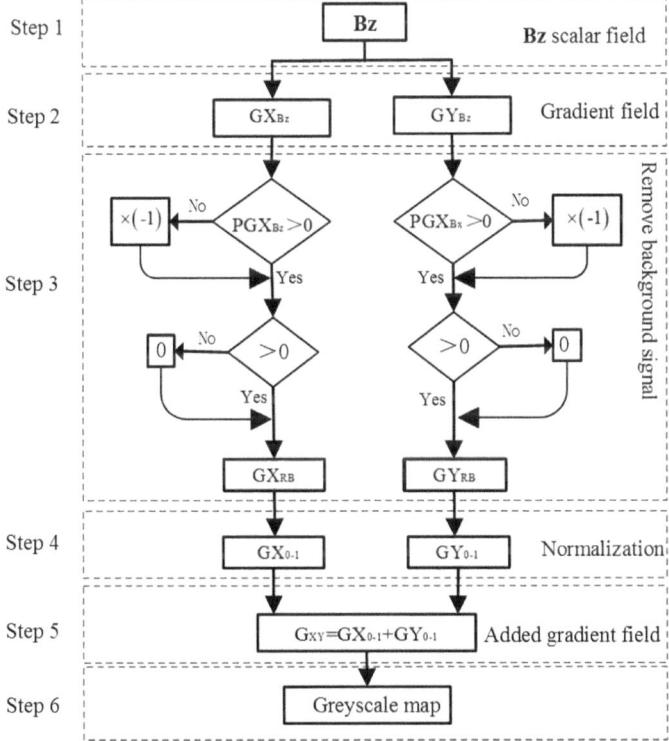

Fig. 6 Double gradient fusion algorithm

- Step 2. Get the double gradient field of the **Bz**. Calculate the **X** direction gradient of each row of the **Bz** ($\partial a_{in}/\partial x$, i = 1, 2, 3…$m$.), forming the **X** direction gradient

$$\text{field } \mathbf{GX_{Bz}} = \begin{bmatrix} \partial xa_{11} \ \partial xa_{12} \cdots \partial xa_{1n} \\ \partial xa_{21} \ \partial xa_{22} \cdots \partial xa_{2n} \\ \partial xa_{31} \ \partial xa_{32} \cdots \partial xa_{3n} \\ \cdots \cdots \cdots \\ \partial xa_{m1} \ \partial xa_{m2} \cdots \partial xa_{mn} \end{bmatrix}.$$

At the same time, Calculate the **Y** direction gradient of each column of the **Bz** ($\partial a_{mj}\big/\partial y, j$ = 1, 2, 3…n.), forming the **Y** direction gradient field $\mathbf{GY_{Bz}} =$

$$\begin{bmatrix} \partial ya_{11} \ \partial ya_{12} \cdots \partial ya_{1n} \\ \partial ya_{21} \ \partial ya_{22} \cdots \partial ya_{2n} \\ \partial ya_{31} \ \partial ya_{32} \cdots \partial ya_{3n} \\ \cdots \cdots \cdots \\ \partial ya_{m1} \ \partial ya_{m2} \cdots \partial ya_{mn} \end{bmatrix}.$$

- Step 3. Remove the background signal of the gradient fields. Find the extreme value of the $\mathbf{GX_{Bz}}$ ($\mathbf{PGX_{Bz}}$) and the extreme value of $\mathbf{GY_{Bz}}$ ($\mathbf{PGX_{Bz}}$). If the extreme value is greater than 0, the negative value in the direction gradient field is removed. If the extreme value is less than 0, the elements in the gradient field multiply − 1 and then the negative value is removed. The **X** direction and **Y** direction removed background are called $\mathbf{GX_{RB}}$ and $\mathbf{GX_{RB}}$ respectively.
- Step 4. Get the normalized gradient field. The $\mathbf{GX_{RB}}$ is normalized to 0–1 forming the $\mathbf{GX_{0\text{-}1}}$. The maximum value of $\mathbf{GX_{0\text{-}1}}$ is 1 and the minimum value of $\mathbf{GX_{0\text{-}1}}$ is 0. The $\mathbf{GY_{RB}}$ is processed in the same way to get the $\mathbf{GY_{0\text{-}1}}$.
- Step 5. Get the fusion gradient field. The fusion gradient field is calculated by the equation: $\mathbf{G_{XY}} = \mathbf{GX_{0\text{-}1}} + \mathbf{GY_{0\text{-}1}}$. The $\mathbf{G_{XY}}$ is plotted with position coordinates forming the fusion gradient field image.
- Step 6. Get the greyscale map. The fusion gradient field image is converted to the greyscale map to achieve visual detection and evaluation of the irregular crack.

4 Experiments

4.1 Testing System

The diagram of the ACFM-MFL testing system is shown in Fig. 7a. The enameled wires are wound on the U-shaped yoke with 500 turns as the excitation coil. The signal generator produces a sine signal with the frequency 1000 Hz and the voltage amplitude 10 V, which is loaded on the excitation coil. The excitation coil excites magnetic field and induces uniform current field in the steel specimen. The tunnel magneto resistance (TMR) chips (Type: TMR2303, Sensitivity 3 mV/Oe, made in MULTI DIMENSION, China.) is located in the probe bottom shell (Thickness 1.0 mm) for

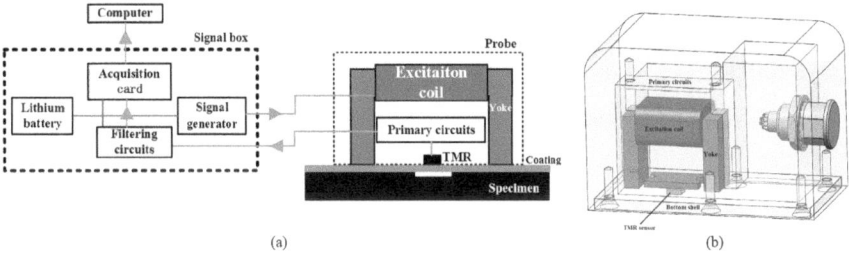

Fig. 7 ACFM-MFL testing system. **a** Diagram of system. **b** Internal structure of probe

the measurement of the distorted magnetic field, as shown in Fig. 7b. The measured magnetic signal is amplified 100 times by the AD620 chip in the primary circuits. The amplified signal is smoothed by the filtering circuits and then captured by the acquisition card. In the end, the signal is processed by the software in the computer.

4.2 Testing Experiments

The specimen is a Q235 steel plate with an irregular crack, as shown in Fig. 8a. The irregular crack consists of 4 cracks, whose angle is $0°$, $30°$, $60°$ and $90°$. The length of the crack is 20 mm and the depth is 2 mm and the width is 0.8 mm. The polyethylene (PE) coating (Thickness 2 mm) is fixed above the irregular crack. The probe is driven by a 3-axial scanner to scan the coating area (90×90 mm^2) to measure the distorted magnetic field, as shown in Fig. 8b. The scan path is a rectangular grid and the step size is 0.5 mm in the **X** direction and the **Y** direction.

The **Bx** and **Bz** signals are picked up at each location to form the scalar field, as shown in Fig. 9. In the **Bx** scalar field, there are peaks at the $90°$ crack and $60°$ crack.

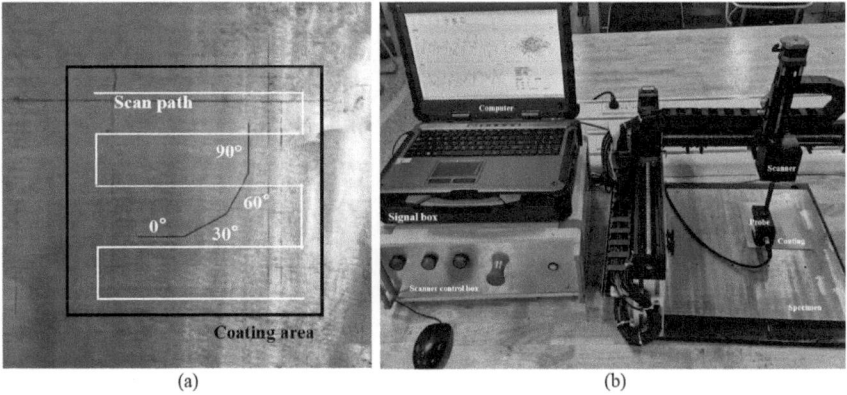

Fig. 8 **a** Photo of specimen with irregular crack (partial view). **b** Photo of testing system

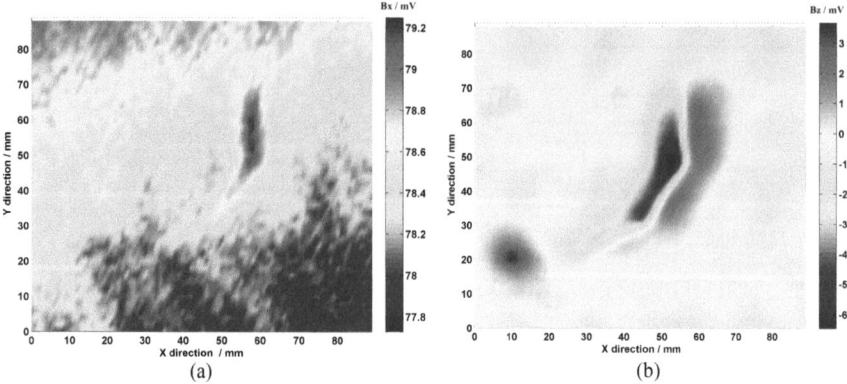

Fig. 9 Testing results of irregular crack. **a Bx** scalar field. **b Bz** scalar field

In the **Bz** scalar field, there are troughs and peaks at each side of the irregular crack. The gathered current field produces peak in the **Bz** at the tips of $0°$ crack. The leaked magnetic field produces obviously peak and trough in the **Bz** at each side of the $90°$ crack. The **Bz** scalar field is more similar to the morphology of the irregular crack, which is in accord with the simulation results. Thus the **Bz** scalar field is set as the characteristic signal to achieve visual evaluation of the irregular crack.

4.3   Verification of Double Gradient Fusion Algorithm

The **Bz** scalar field is processed by the double gradient fusion algorithm step by step. The **X** direction and **Y** direction gradient field of the **Bz** are shown in Fig. 10a, b respectively. Then the background signals of the GX_{Bz} and GY_{BZ} are removed to get the GX_{RB} and GX_{RB}, as shown in Fig. 10c, d respectively. Next, the GX_{RB} and GY_{RB} are normalized to GX_{0-1} and GY_{0-1}, as shown in Fig. 10e, f respectively. The $30°$ crack, $60°$ crack and $90°$ crack can be seen in Fig. 10e because of the magnetic leakage effect. Meanwhile, the $0°$ crack, $30°$ crack and $60°$ crack can be seen in Fig. 10f due to the current perturbation effect. The experimental results are consistent with the theory and simulation results.

In the end, the GX_{0-1} adds the GY_{0-1} to form the G_{XY}, as shown in Fig. 11a. Figure 11a is converted to the greyscale map, as shown in Fig. 11b. The morphology of the irregular crack is shown obviously, which accords with the true shape of the irregular crack in the steel. To achieve visual evaluation of the irregular crack, the tips of the irregular crack are picked up in the greyscale map. The irregular crack is reconstructed and plotted by the location of the tips, as shown in Fig. 11c. The reconstructed image is in accord with the full morphology of the irregular crack.

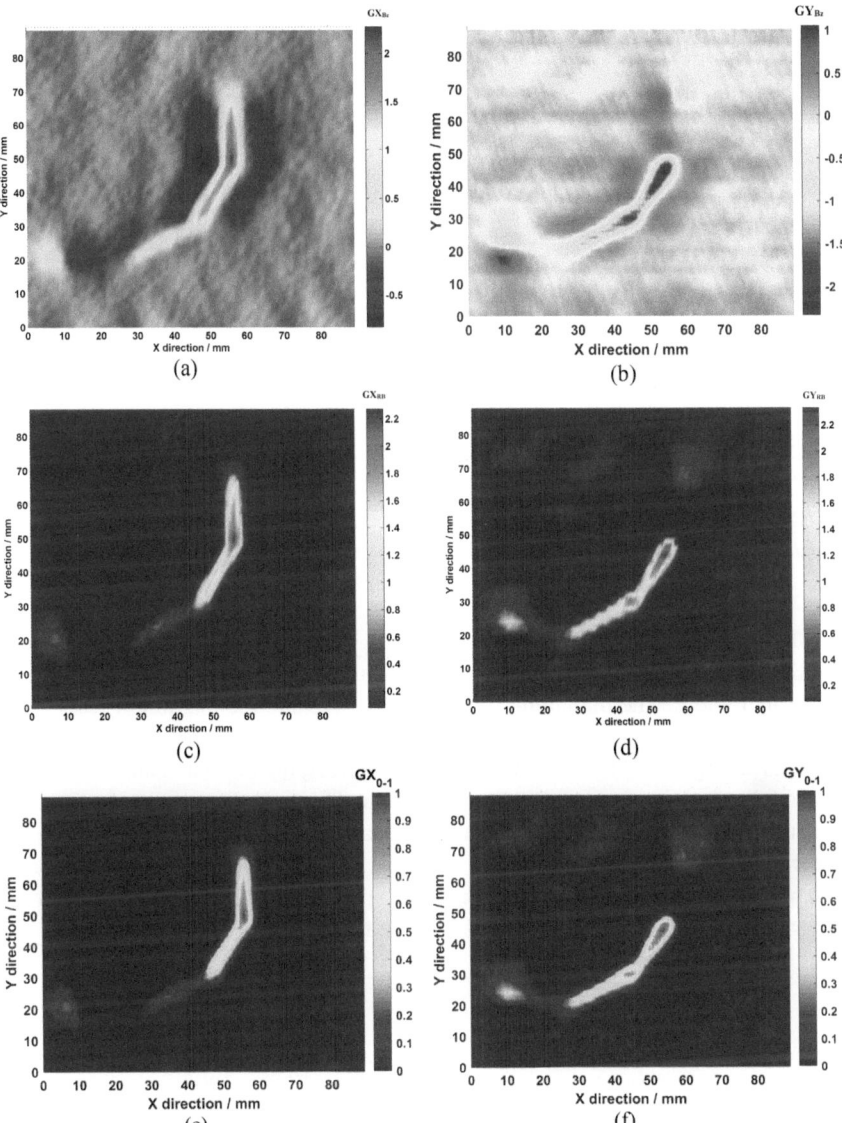

Fig. 10 Double gradient fusion algorithm for experimental **Bz** scalar field. **a** X direction gradient field GX_{Bz}. **b** Y direction gradient field GY_{Bz}. **c** Removed background signal GX_{RB}. **d** Removed background signal GY_{RB}. **e** Normalized $GX_{0\text{-}1}$. **f** Normalized $GY_{0\text{-}1}$

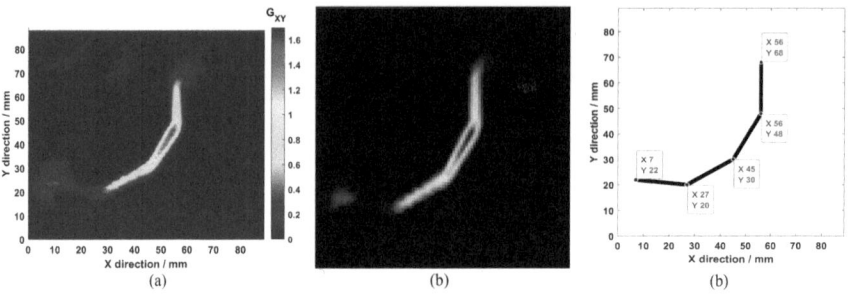

Fig. 11 Visual evaluation results of irregular crack. **a** Fusion gradient field G_{XY}. **b** Greyscale map. **b** Location of crack tips

Table 2 Evaluated results of irregular crack

Name	0° crack	30° crack	60° crack	90° crack
Length/mm	20.1	20.6	21.1	20.0
Angle	5.7°	29.1°	58.6°	90.0°

The length and angle of the irregular crack can be evaluated by the location of the tips, as given in Eqs. (2) and (3). The evaluated results of the irregular crack is given in Table 2. The maximum angle error is 5.7° and the maximum length error is 1.1 mm whose maximum relative error is 5.5%. Because the magnetic leakage effect is stronger than the current perturbation effect, the evaluated angle and length of the 90° crack is more precise than that of the 0° crack. The results show that the irregular crack can be evaluated visually and accurately by the location of the tips. We can make a conclusion that the irregular crack in the steel under the coating can be reconstructed and evaluated visually by the double gradient fusion algorithm using the composite ACFM-MFL testing method.

$$Length = \sqrt{(X_1 - X_2)^2 + (Y_1 - Y_2)^2} \tag{2}$$

$$Angle = artan(|Y1 - Y2|/|X1 - X2|) \times 180/\pi \tag{3}$$

5   Conclusion and Further Work

The double gradient fusion algorithm is presented to achieve visual evaluation of the irregular crack using the composite ACFM-MFL testing method. The leaked primary magnetic field and distorted secondary magnetic field are analyzed by the FEM model. The **Bz** is defined as the characteristic signal to reconstruct the morphology of the irregular crack. The double gradient fusion algorithm is presented to process the **Bz** scalar field for the visual evaluation of the irregular crack. The ACFM-MFL testing system is developed to detect the irregular crack in the steel under the coatings. The efficiency of the double gradient fusion algorithm is verified by the experimental **Bz** scalar field. The results show that the composite ACFM-MFL testing method is sensitive to any part of the irregular crack due to the combination of the magnetic leakage effect and current field perturbation effect. The whole morphology of the irregular is reconstructed accurately by the double gradient fusion algorithm. As a results, the irregular crack is evaluated visually by the greyscale map of the crack morphology. The maximum evaluated angle error is $5.7°$ and the maximum evaluated length error is 1.1 mm in the experiment. The double gradient fusion algorithm and composite ACFM-MFL testing method can be used in the detection and evaluation of any direction cracks in the steel. The morphology of various cracks can be shown visually without removing the coatings. Further work will focus on the evaluation of the profile of irregular cracks, the development of high resolution sensor array for online visual evaluation of crack under the coatings.

References

1. Corcoran J, Hooper P, Davies C, Nagy PB, Cawley P (2016) Creep strain measurement using a potential drop technique. Int J Mech Sci 110:190–200
2. Cao W et al (2019) Microstructural material characterization of hypervelocity-impact-induced pitting damage. Int J Mech Sci 163:105097
3. Lofrano E, Paolone A, Ruta G (2020) Dynamic damage identification using complex mode shapes. Struct Control Health Monit Sci 27:e2632
4. Gao X, Wang Y, Zhang Z, Tan J, Cheng L, Yang W (2019) Conductive nano-carbon coating on silica by pyrolysis of polyethylene. Mater Lett 255:126567
5. Earls CJ (2013) Bayesian inference of hidden corrosion in steel bridge connections: non-contact and sparse contact approaches. Mech Syst Signal Process 41(1–2):420–432
6. Chen Q, Huang Y, Weng X, Liu W (2021) Curve-based crack detection using crack information gain. Struct Control Health 4(8):1–27
7. Khan A, Ali SSA, Meriaudeau F, Malik AS, Soon LS, Seng TN (2017) Visual feedback-based heading control of autonomous underwater vehicle for pipeline corrosion inspection. Int J Adv Robot Syst 3:177–189
8. Andrushia D, Anand N, Arulraj P (2020) A novel approach for thermal crack detection and quantification in structural concrete using ripplet transform. Struct Control Health 11:1–20
9. Lee JS, Hwang SH, Choi IY, Choi Y (2020) Estimation of crack width based on shape-sensitive kernels and semantic segmentation. Struct Control Health Monitor 27(4):1–21
10. Zhao J, Li W, Ge J (2020) Coiled tubing wall thickness evaluation system using pulsed alternating current field measurement technique. IEEE Sensors J 18:10495–10501
11. Xu Y, Yang Y, Wu Y (2020) Eddy current testing of metal cracks using spin Hall magnetoresistance sensor and machine learning. IEEE Sensors J 20(18):10502–10510
12. Xie SA et al (2020) Features extraction and discussion in a novel frequencyband-selecting pulsed eddy current testing method for the detection of a certain depth range of defects. NDT E Int 111:102211
13. Wang C, Fan M, Cao B, Ye B, Li W (2018) Novel noncontact eddy current measurement of electrical conductivity. IEEE Sensors J 18(22):9352–9359

14. Li X, Gao B, Woo WL, Tian GY, Qiu X, Gu L (2017) Quantitative surface crack evaluation based on eddy current pulsed thermography. IEEE Sensors J 17(2):412–421
15. Rowshandel H, Nicholson GL, Shen JL, Davis CL (2018) Characterisation of clustered cracks using an ACFM sensor and application of an artificial neural network. NDT E Int 98:80–88
16. Raine A (2002) Cost benefit applications using the alternating current field measurement testing technique. Mater Eval 1:49–52
17. Dover WD, Collins R, Michael DH, Thompson RB (1986) The use of AC-field measurements for crack detection and sizing in air and underwater. Phil Trans Roy Soc Lond Ser A Math Phys Sci 1554:271–283
18. Lewis AM, Michael DH, Lugg MC, Collins R (1988) Thin-skin electromagnetic fields around surface-breaking cracks in metals. J Appl Phys 8:3777–3784
19. Pasadas DJ, Ribeiro AL, Rocha T, Ramos HG (2016) 2D surface defect images applying Tikhonov regularized inversion and ECT. NDT E Int 80:48–57
20. Feng Y, Zhang L, Zheng W (2018) Simulation analysis and experimental study of an alternating current field measurement probe for pipeline inner inspection. NDT E Int: 123–129
21. Rowshandel H, Nicholson GL, Shen JL, Davis CL (2018) Characterisation of clustered cracks using an ACFM sensor and application of an artificial neural network. NDT E Int 98:80–88
22. Shen JL, Zhou L, Rowshandel H, Nicholson GL, Davis CL (2015) Determining the propagation angle for non-vertical surface-breaking cracks and its effect on crack sizing using an ACFM sensor. Meas Sci Technol 26(11):115604
23. Li W, Yuan X, Chen G, Ge J, Yin X, Li K (2016) High sensitivity rotating alternating current field measurement for arbitrary-angle underwater cracks. NDT E Int 79:123–131
24. Zhang N, Ye C, Peng L, Tao Y (2019) Novel array eddy current sensor with three-phase excitation. IEEE Sensors J 19(18):7896–7905
25. Wang Y, Ye C, Wang M (2019) Synthetic magnetic field imaging with triangle excitation coil for inspection of any orientation defect. IEEE Trans Instrum Meas 69(2):533–541
26. Zhang N, Ye C, Peng L, Tao Y (2020) Eddy current probe with three-phase excitation and integrated array tunnel magnetoresistance sensors. IEEE Trans Ind Electron 68(6):5325–5336
27. Ge J et al (2017) Analysis of signals for inclined crack detection through alternating current field measurement with a U-shaped probe. Insight Non-Destr Test Condition Monitor 59(3):121–128
28. Wu D, Liu Z, Wang X, Su L (2017) Composite magnetic flux leakage detection method for pipelines using alternating magnetic field excitation. NDT E Int 91(1):140–155
29. Gong Q, Zhu L, Wang Y, Yu Z (2021) Automatic subway tunnel crack detection system based on line scan camera. Struct Control Health 8:1–22
30. Gao Y, Su Y, Li Q, Li H, Li J (2020) Single image dehazing via self-constructing image fusion. Signal Process 167:107284
31. Wu L, Liu Y, Liu N, Zhang C (2016) High-resolution images based on directional fusion of gradient. Comput Vis Media 4(1):31–43
32. Luo L, Feng MQ, Wu J, Bi L (2021) A vision-based surface displacement/strain measurement technique based on robust edge-enhanced transform and algorithms for high spatial resolution. Struct Control Health Monit 28(9):1–31

Design and Experiment Research of Oblique Crack Detection System for Rail Tread Based on ACFM Technique

1 Introduction

Railway transportation, as one of the main transportation methods in China, plays an important role in the entire transportation system. In the past decade, China has implemented six large-scale train speed increases, which have greatly improved railway transportation capacity while also presenting challenges for safe train operations [1, 2]. Due to the action of wheel-rail cyclic loading, the rail tread is prone to micro-cracks. These cracks are typical rolling contact fatigue cracks, with the crack's horizontal angle usually at 45° to the direction of travel. Initially, they form at an angle of 10–15° perpendicular to the rail tread surface. After reaching a certain depth, these cracks rapidly propagate towards the interior of the rail head at a larger angle, leading to surface stripping, spalling, or even rail breakage. This poses a serious threat to train safety [3–5]. Therefore, conducting inspections for oblique cracks on the rail tread surface is crucial for ensuring safe railway operations and preventing accidents.

At present, the commonly employed non-destructive testing techniques for rail tread surface defects include ultrasonic testing, magnetic flux leakage testing, eddy current testing, and visual inspection. However, the existing detection methods still have limitations in effectively identifying surface defects on the rail tread surface [6–11]. Ultrasonic testing has been widely used for the detection and assessment of internal defects in rails, but it requires a coupling agent during testing and is subject to blind spots in surface and near-surface inspection [12]. Magnetic flux leakage testing requires magnetization of the rail, and it is not sensitive to small open surface and near-surface cracks [13]. Eddy current testing can effectively detect surface and near-surface defects, but the detection results are greatly influenced by the probe lift-off [14]. Visual inspection techniques can accurately identify large-scale defects and corrosion damage on rail tread surfaces. However, it is challenging to detect micro-cracks on the surface and near-surface areas, and it cannot provide information about the depth of the cracks [15].

W. Li et al., *Alternating Current Field Measurement Technique for Detection and Measurement of Cracks in Structures*, https://doi.org/10.1007/978-981-97-7255-1_4

Compared to the aforementioned non-destructive testing techniques, Alternating Current Field Measurement (ACFM) technology combines the uniform current field of alternating current voltage drop detection and the non-contact detection of eddy current testing. It has the advantages of precise mathematical modeling, insensitivity to lift-off, simultaneous qualitative and quantitative assessment, and non-retrospective detection [16–18]. Currently, some researchers have applied ACFM technology in the field of rail transportation. Nicholson et al. [19] utilized finite element software to establish a simulation model for rolling fatigue crack ACFM, investigating the relationship between rolling fatigue cracks and characteristic signals, and conducting experimental verification. Rowshandel et al. [20] combined artificial neural networks with ACFM technology to achieve size assessment of fatigue cracks in steel rails and wheel-rail systems. Shen et al. [21] investigated the relationship between complex-shaped rolling fatigue cracks and ACFM characteristic signals, and proposed a crack profile reconstruction method. The University of Birmingham in the UK has established a high-speed rail measurement platform, using ACFM detection instruments and pencil probes from TSC company, and employing B-spline functions to reconstruct the response curve of characteristic signals at high speed, achieving the detection of rail defects at 121.5 km/h [22]. Ge et al. [23] added a "slipper" structure to the ACFM probe to avoid collision with steel rails and switches, analyzed the impact of speed on detection signals, and achieved high-speed detection of surface cracks on steel rails. From the above, it can be seen that current research on the detection of rail surface defects mainly focuses on the processing of characteristic signals. The experimental systems used generally employ common AC electromagnetic field detection probes and equipment, lacking specially designed detection systems tailored to the structure and defect distribution characteristics of rail surfaces. This limitation has constrained the improvement of rail surface defect detection effectiveness.

The principle of AC electromagnetic field detection for oblique cracks on rail surfaces is shown in Fig. 1. When a sinusoidal AC signal is applied to the excitation coil, the coil induces a uniform induced current on the rail surface through a U-shaped magnetic core. If there is a crack on the rail surface, the uniform induced current bypasses the crack from both sides and the bottom, causing spatial magnetic field distortion. The magnetic sensor arranged above the rail surface picks up the distorted magnetic field signal, enabling the detection and quantification of cracks.

This paper focuses on the structural characteristics of rail surfaces and the detection requirements for oblique cracks, applying AC electromagnetic field detection technology to detect oblique cracks on rail surfaces. By establishing a finite element simulation model of oblique cracks on rail surfaces using ACFM, the disturbance mechanism of oblique cracks on induced electromagnetic fields and the influence of scanning paths on characteristic signals were analyzed. Based on the characteristics of rail surfaces, a rail surface detection probe and inspection scanning frame were designed, and the development of detection instruments and software was completed. Finally, a complete detection system was formed, which achieved effective detection of oblique cracks on rail surfaces.

Fig. 1 ACFM rail surface oblique crack detection principle diagram

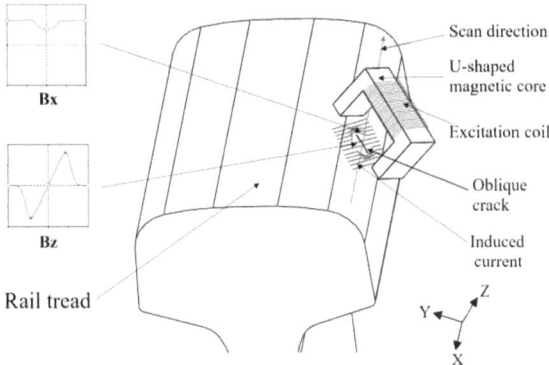

2 Finite Element Method Model

2.1 Simulation Model

Based on the characteristics of oblique cracks on rail surfaces, this paper establishes an ACFM simulation model for oblique cracks on rail surfaces using the finite element simulation software COMSOL. Considering the trade-off between solution time and accuracy, the simulation model is simplified as shown in Fig. 2. The model mainly includes a simulated rail specimen, cracks, and air. The simulation analysis is conducted by applying an excitation current density of 100 A/m with a frequency of 1 kHz perpendicular to the crack length direction. The simulated rail specimen is made of carbon steel. The dimensional parameters and material properties of the simulation model are summarized in Tables 1 and 2, respectively.

Since cracks on rail surfaces usually originate from the gauge corner, the horizontal angle of the crack is typically 45° with respect to the direction of travel, and it forms a certain angle with the vertical direction and the rail surface. Therefore, in the simulation model, the horizontal angle of the crack is set to 45°, the vertical angle is set to 30°, the length of the crack is 12 mm, and the depth is 5 mm, as shown in Fig. 3. The established finite element model is used to simulate and analyze the crack, and the distribution of induced currents around the crack is extracted as shown in Fig. 4. From the figure, it can be observed that in the absence of a crack, the surface induced current is uniformly distributed and perpendicular to the crack direction. The cross-sectional induced current follows the skin effect distribution pattern, with higher current density on the crack surface and gradually decreasing in the depth direction. When a crack exists, the induced current is disturbed, and the surface current bypasses the two endpoints of the crack, with one endpoint in a clockwise direction and the other endpoint in a counterclockwise direction. The cross-sectional induced current bypasses the bottom of the crack. In order to simulate the scanning process of the detection probe and analyze the response characteristics of the characteristic signals at the center position of the crack, the magnetic flux density values at a position

Fig. 2 Finite element simulation model of ACFM for rail surface oblique cracks. **a** Overall view of the simulation model. **b** Local view of the oblique crack region

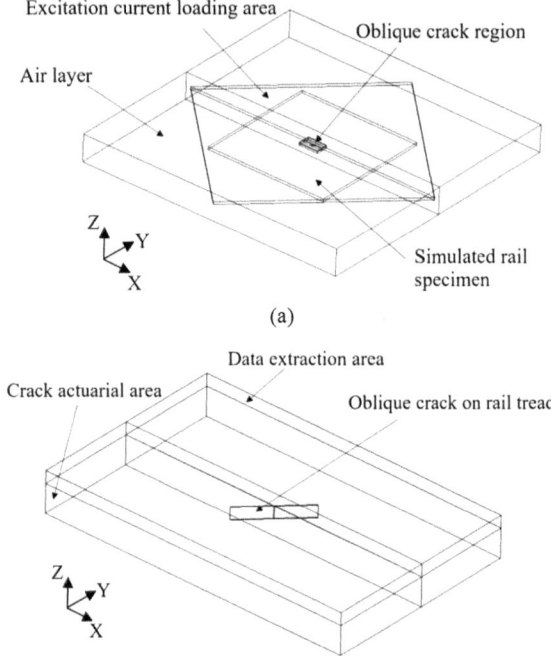

(a)

(b)

Table 1 Model dimensions

Model	Length/mm	Width/mm	Depth/mm
Air layer	1100	1100	100
Simulated steel rail specimens	500	500	10
Excitation current loading region	750	750	10
Crack evaluation region	80	50	7
Data extraction region	80	50	3

Table 2 Model material parameter properties

Model	Model material	Electrical conductivity/S m^{-1}	Relative permeability	Relative dielectric constant
Air layer	Air	50	1	1
Excitation current loading region	Air	50	1	1
Simulated steel rail specimens	Steel of the rail	5×106	150	1
Inclined crack	Ai	50	1	1

Fig. 3 Graphical representation of oblique cracks on the rail surface

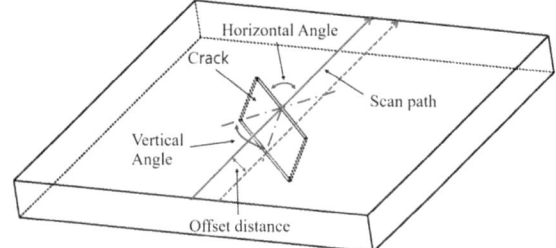

0.5 mm above the crack are extracted along the scanning path, as shown in Fig. 5. In the figure, Bx represents the magnetic flux density component along the scanning path, and Bz represents the magnetic flux density component normal to the rail surface. From Fig. 5, it can be seen that the magnetic flux density value above the defect-free rail surface remains constant, while at the crack location, the Bx signal shows a trough, and the Bz signal exhibits peaks and troughs sequentially.

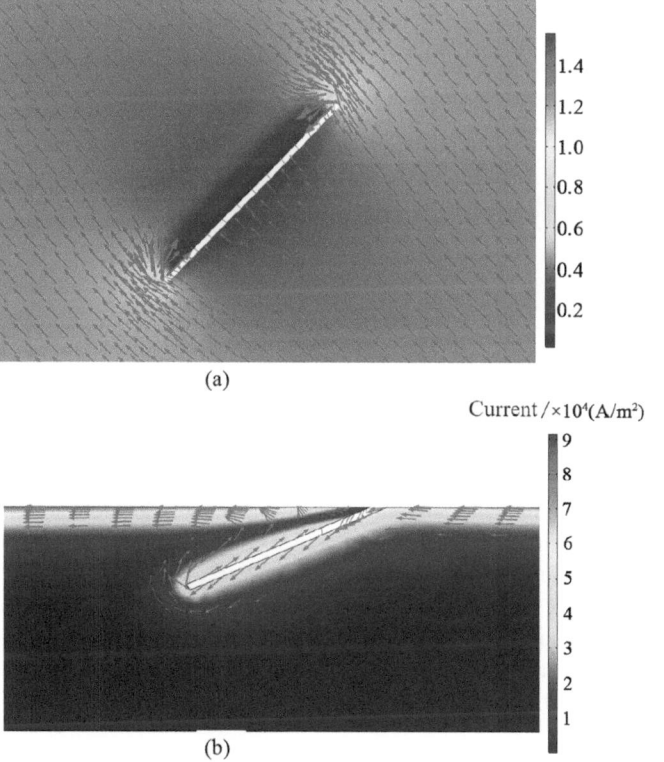

Fig. 4 Induced current. **a** Surface current. **b** Internal current

Fig. 5 Feature signal

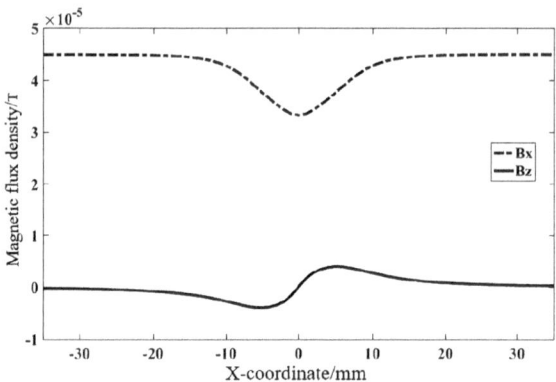

2.2 The Influence of the Scanning Path on the Characteristic Signals

During the process of rail surface inspection, the position of the crack is unknown, so it is necessary to explore the influence of the scanning path on the characteristic signals. The magnetic field signals from different scanning paths are extracted as shown in Fig. 6, with the legend distance representing the perpendicular distance from the center of the skewed crack to the scanning path; when the distance is 0 mm, the scanning path passes through the center of the crack. In order to more intuitively express the spatial magnetic field variation above the crack, the Bx magnetic field signal is uniformly baseline processed, and the uniformly baselined Bx signal is defined as the NBx signal, as shown in Eq. 1:

$$NBx = Bx/Bx_0 \times 100\% \tag{1}$$

In the equation, Bx0 represents the baseline magnetic field signal value at the location without a crack.

From Fig. 6a, it can be observed that when the offset distance is 0 mm, the NBx signal is a trough signal; as the vertical distance gradually increases to 8 mm, the trough decreases gradually, and a new peak appears on the right side of the trough, reaching its maximum value at 7 mm. In Fig. 6b, as the scanning path deviates from the crack center, the left trough of the Bz signal gradually decreases, while the right peak continuously increases, reaching its maximum value when deviating by 6 mm. This is due to the induced current bypassing the crack from both ends, resulting in the accumulation and increase in current density at the ends of the crack, thereby increasing the spatial disturbance of the magnetic field signal.

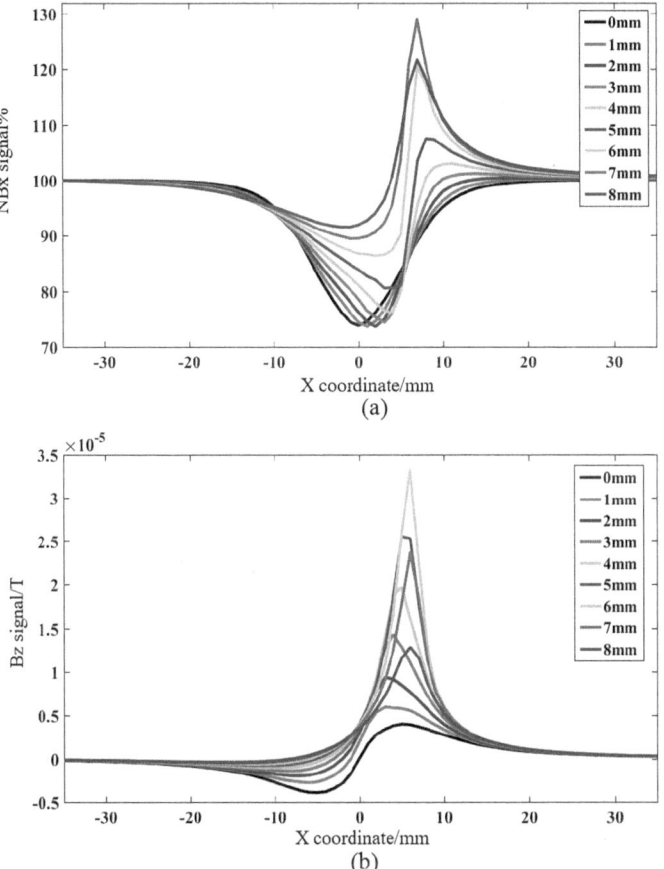

Fig. 6 The influence of scanning paths on the magnetic field signals **a** Bx magnetic field signal. **b** Bz magnetic field signal ion

In order to provide a more precise description of the magnetic field characteristic signals of the crack, the distortion variables ΔBx for the Bx signal and ΔBz for the Bz signal are defined as shown in Eqs. 2 and 3, respectively.

$$\Delta Bx = NBx_{\max} - NBx_{\min} \tag{2}$$

$$\Delta Bz = Bz_{\max} - Bz_{\min} \tag{3}$$

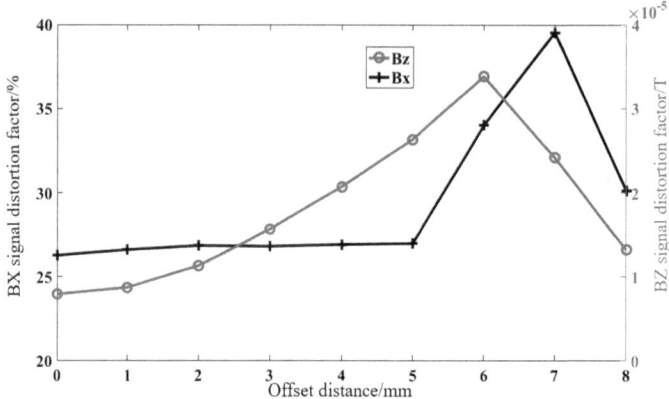

Fig. 7 Influence of scanning path on feature signal distortion

In the equations, NBxmax and NBxmin represent the maximum and minimum values of the NBx signal, while Bzmax and Bzmin represent the maximum and minimum values of the Bz signal, typically corresponding to the peak and trough values of the Bz signal.

Figure 7 shows the variation curves of the distortion variables of the Bx and Bz signals as the scanning path deviates from the crack center. From the graph, it can be observed that the distortion variable of the Bx signal remains relatively small within the range of 5 mm, continuously increases between 5 and 7 mm, reaching its maximum value at 7 mm. This is because the magnetic leakage field rapidly increases at the endpoint positions of the crack. Beyond 7 mm, the Bx distortion variable starts to decrease. On the other hand, the distortion variable of the Bz signal continuously increases from 0 to 6 mm and then decreases from 6 to 8 mm. For different scanning paths, the distortion variables of the characteristic signals vary. The distortion variable is smallest at the crack center position and largest near the crack endpoints. Therefore, the extremum points of the characteristic signals can be utilized to measure the length of the crack.

3 Testing System

3.1 Probe Design

To achieve rapid detection of the entire rail tread surface, a probe for detecting oblique cracks in rail tread surface was designed. First, the magnetic core was simulated and analyzed as shown in Fig. 8. A custom-shaped magnetic core was designed based on the special structure of the rail gauge angle. The two legs of the magnetic core have different lengths, and the bottom is designed with angled surfaces of different

Fig. 8 Current distribution
on the rail surface

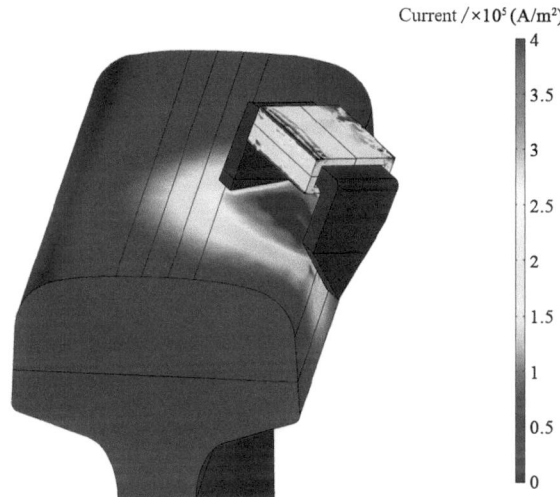

Current /×10⁵ (A/m²)

angles to match the rail surface, which minimizes the air gap. While ensuring that the induced magnetic field covers the entire rail tread surface area, the design achieves a higher induction current intensity and a more uniform magnetic flux density on the rail tread surface.

Taking into account the detection accuracy of the array probe and the sensitivity of the magnetic field sensor, TMR (Tunneling Magnetoresistance) sensors were chosen as the array-type magnetic field sensors. The spacing between the sensors was set to 5 mm. Oblique cracks on the rail tread surface develop from the rail gauge angle towards the rail tread surface and typically have a length of around 6–30 mm, forming a 45° angle with the direction of travel. Therefore, setting up 5 array TMR sensors is sufficient to cover the oblique cracks. The distance between adjacent sensors is set to 5 mm. The distribution schematic of sensors 1–5 at the rail gauge angle is shown in Fig. 9.

According to the distribution characteristics of cracks on the rail tread surface, a rail-shaped array detection probe was designed as shown in Fig. 10. The probe utilizes a specially shaped magnetic core that matches the rail gauge angle and is inclined at a 45-degree angle. The induced current direction is perpendicular to the oblique cracks. The outer shell of the array probe is designed to closely fit with the rail, and the array sensors are located at the bottom of the probe, maintaining a small separation distance from the rail surface.

Fig. 9 Distribution of array sensors on the rail surface

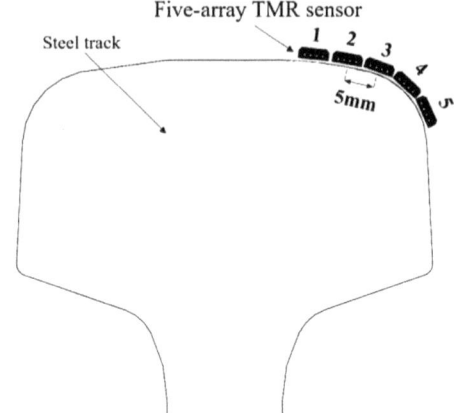

Fig. 10 Rail inspection array probes. **a** Design diagram of the conformal array probe. **b** Photo of the conformal array probe

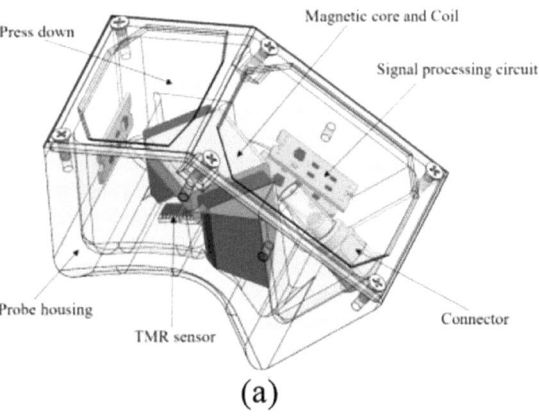

(a)

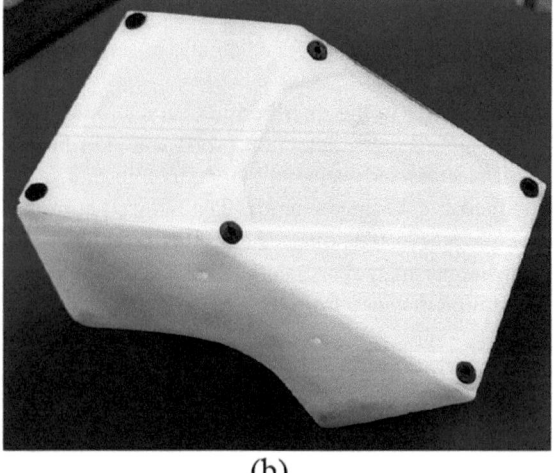

(b)

3.2 Detection System

To meet the requirements of stability, portability, and efficiency in rail inspection, an ACFM (Alternating Current Field Measurement) detection system for rail tread surface oblique cracks is designed as shown in Fig. 11. The overall solution consists of a hardware system and a software system. The hardware system includes a detection probe, a rail inspection scanning frame, and a rail crack ACFM detection instrument. The rail inspection scanning frame is made of aluminum profiles and is equipped with wheels at the bottom to ensure stability during the detection process. The probe fixture on the scanning frame enables the fixation and compression of the detection probe, reducing probe vibration and improving detection stability. The probe fixture can accommodate different types of detection probes. The rail oblique crack ACFM detection instrument is fixed on the bracket of the scanning frame. It mainly consists of a power module, a signal generator, connectors, effective value processing circuitry, data acquisition card, and an industrial-grade PC. The highly integrated components enable miniaturization and integration of the detection instrument. The instrument is connected to the detection probe and encoder through connectors and signal cables. It outputs excitation signals from the signal generator while receiving detection signals from the probe and position signals from the encoder. The software system is developed using LabVIEW and MATLAB, enabling data display and analysis of the rail tread surface oblique crack detection signals.

4 Experimental Research

4.1 Different Depth Oblique Crack Detection Experiments

In order to analyze the influence of crack depth on the characteristic signals, detection experiments were conducted on rail specimens with different crack depths. The excitation frequency of the detection instrument was set at 1 kHz, and the material of the specimens was 60 kg/m rail. The crack depths were 1, 3, 5, and 7 mm, with a length of 18 mm and a width of 0.3 mm. The cracks were inclined at a horizontal angle of 45° and a vertical expansion angle of 50°, as shown in Fig. 12.

The detection probe was used to scan different depth cracks, and C-scan images were generated based on the position relationship and the characteristic signal Bz of different sensors, as shown in Fig. 13. The position at a lateral distance of 0 mm represents the center position of sensor 3, and the centers of adjacent sensors have a lateral distance difference of 5 mm. From the image, it can be observed that the cracks cause distortion in the induced magnetic field signal. The red region represents the peak of the Bz signal, and the blue region represents the trough of the Bz signal. The central regions of the red and blue areas represent the two endpoints of the crack. As the crack depth increases, the distortion of the Bz signal also increases. Among them, the 1 mm deep crack has the smallest distortion amplitude, measuring 36

Fig. 11 ACFM detection
system for oblique cracks on
the rail surface. **a** Schematic
diagram of the detection
system. **b** Photo of detection
system

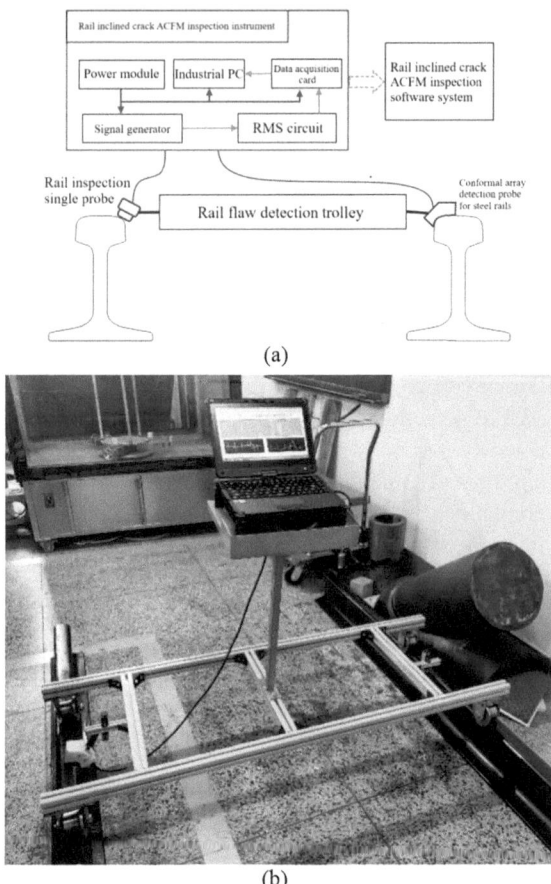

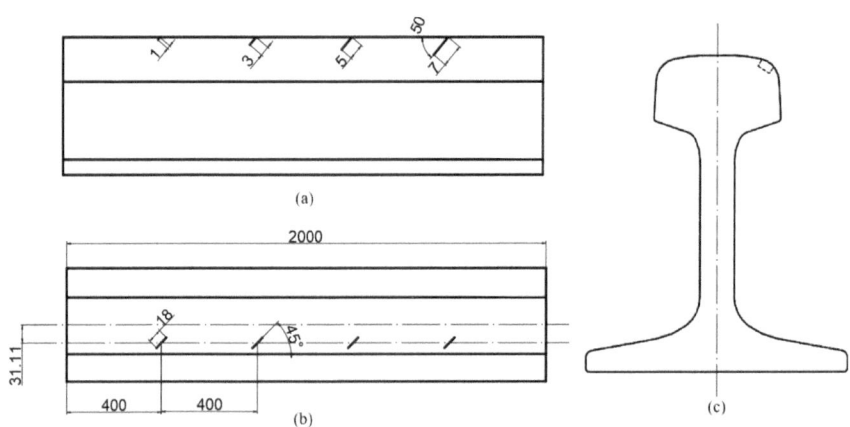

Fig. 12 Schematic diagram of steel rail specimens with cracks at various depths. **a** Front view of
the specimen. **b** Top view of the specimen. **c** Left view of the specimen

mV, while the 7 mm deep crack has the largest distortion amplitude, measuring 209 mV. For different sensor positions, sensor 2 and sensor 5 have the largest distortion amplitudes, indicating that the crack endpoints are close to the scanning path of these two sensors. Additionally, there is a difference in the peak and trough values of the signals from these two channels in the direction of travel, indicating a certain horizontal angle between the crack and the scanning direction. Based on the C-scan image, the preliminary determination of the horizontal angle of the crack is around 45°, with a length of approximately 20 mm, which is close to the actual value of 18 mm. Therefore, it can be concluded that the detection instrument can detect cracks with a minimum depth of less than 1mm and provide preliminary quantification of the crack's horizontal angle and length. This meets the requirements for array imaging and large-scale detection of oblique cracks at rail gauge corners.

In order to analyze the influence of cracks of different depths on the characteristic signals, the characteristic signals along the scanning path at the center of the cracks were extracted, as shown in Fig. 14. Simultaneously, the curves of the distortion amplitudes of the Bx and Bz signals for different crack depths were plotted, as shown in Fig. 15. From the figures, it can be observed that as the crack depth increases, the depth of the Bx signal troughs also increases, but the range of distortion remains relatively constant. The peaks and troughs of the Bz signal show an increasing trend, while the horizontal spacing between them remains relatively constant. This is because as the crack depth increases, the distortion amplitude of the induced electric field increases, resulting in an increase in the distortion amplitude of the induced magnetic field.

4.2 Detection Experiment of Inclined Cracks of Different Lengths

In order to analyze the influence of crack length on the characteristic signals, detection experiments were conducted on rail specimens with cracks of different lengths. The crack lengths were 6, 12, 18, 24, and 30 mm, with a depth of 5 mm and a width of 0.3 mm. The cracks were inclined at a horizontal angle of 45° and a vertical expansion angle of 30°, as shown in Fig. 16.

Using array probes to detect cracks of different lengths separately, the C-scan images of different crack lengths shown in Fig. 17 were obtained. From the images, it can be observed that as the crack length increases, the distance between the peak and trough of the Bz signal becomes larger, approaching different sensors. At the same time, as the crack length increases, the overall distortion of the signal also increases. This is because longer cracks induce a higher current density around their ends, resulting in a stronger magnetic field signal in the Z direction. The length of the crack can be obtained by measuring the distance between the peak and trough of the Bz signal, and through five measurements, the average measurement error of the array probe for crack length is less than 2 mm.

Fig. 13 Different depth crack array probe results. **a** Crack depth of 1 mm. **b** Crack depth of 3 mm. **c** Crack depth of 5 mm. **d** Crack depth of 7 mm

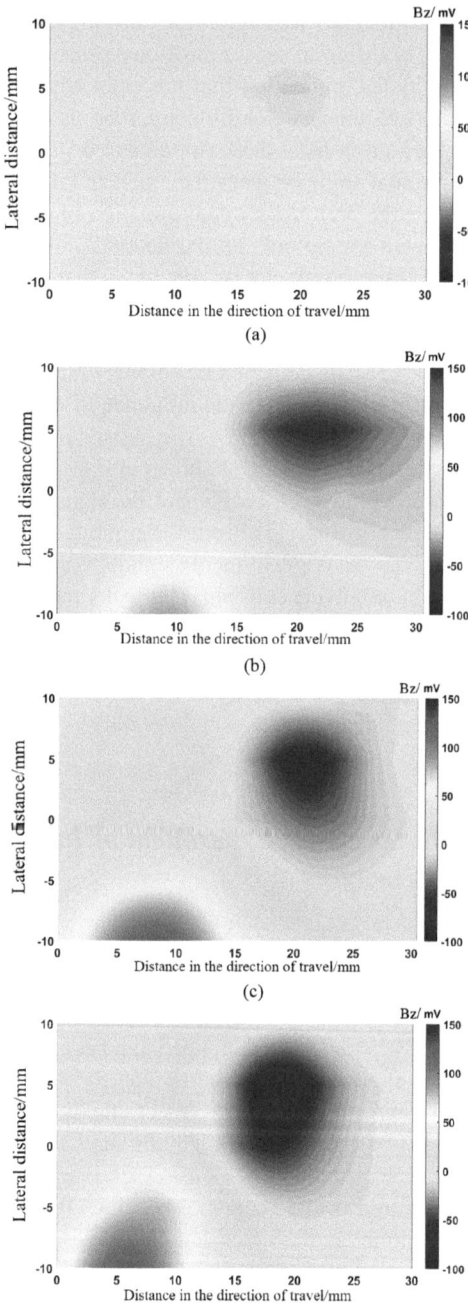

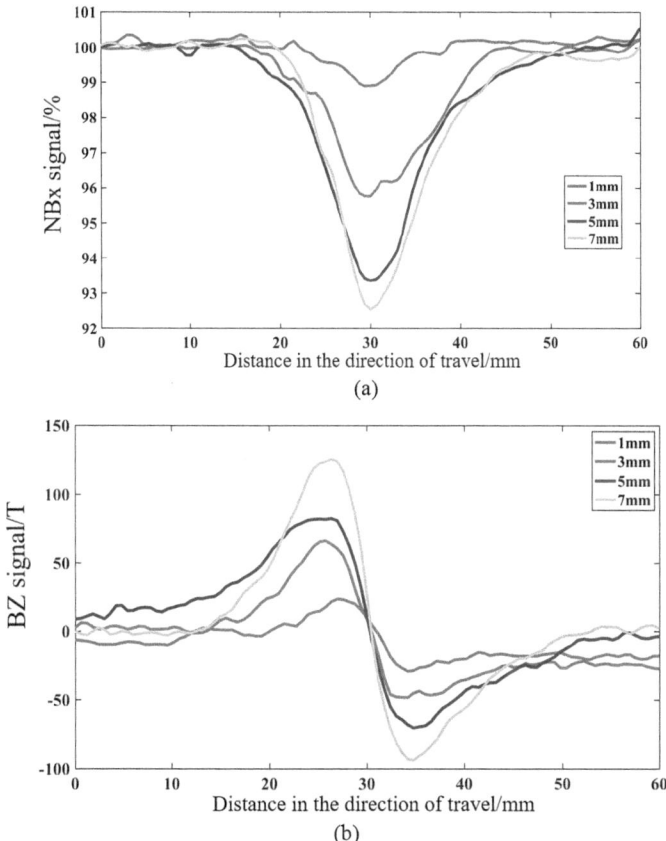

(a)

(b)

Fig. 14 Different depth crack single probe results. **a** BX signal. **b** BZ signal

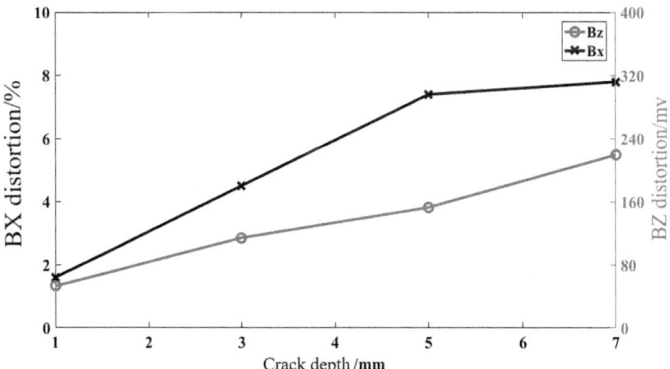

Fig. 15 The impact of crack depth on characteristic signalsl

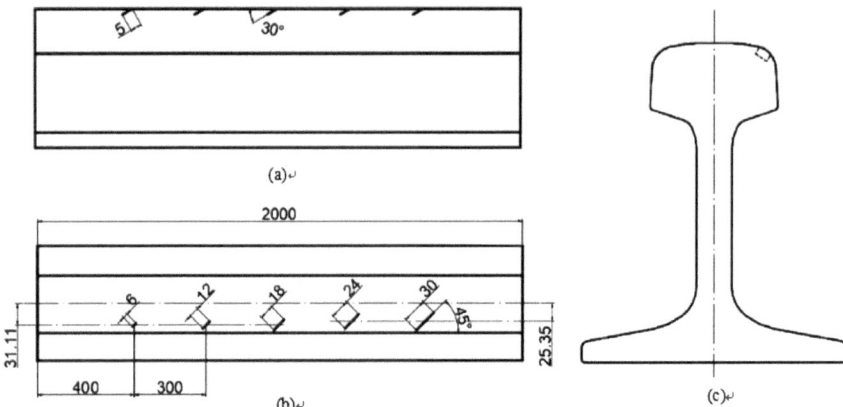

Fig. 16 Schematic diagram of steel rail specimens with cracks of different lengths. **a** Front view of the specimen. **b** Top view of the specimen. **c** Left view of the specimen

In order to analyze the influence of crack length on the characteristic signals, the characteristic signals along the scanning path at the center of the cracks were extracted, as shown in Fig. 18. Simultaneously, the distortion amplitudes of the Bx and Bz signals were plotted, as shown in Fig. 19. From the figures, it can be observed that as the crack length increases, the distortion amplitude of the Bx signal initially increases and then decreases, with an increasing range of distortion. Similarly, the distortion amplitude of the Bz signal also increases and then decreases, with an increasing horizontal spacing between the peaks and troughs. This is because as the crack length increases, the range and density of the induced electric field distortion increase, leading to an increase in the distortion amplitude of the characteristic signals. However, as the crack length continues to increase, the distance between the crack center and the endpoints increases, weakening the influence of the distortion current at the endpoints on the central region, resulting in a decrease in the distortion amplitude of the characteristic signals.

5   Conclusions

This paper focuses on the characteristics of rail surface inspection and designs an ACFM (Alternating Current Field Measurement) system for detecting oblique cracks on the rail surface. Through simulation models, the response mechanism between the distorted electromagnetic field and oblique cracks is investigated. The influence of scanning paths on the characteristic signals is analyzed. Additionally, a detection probe, scanning frame, and detection instrument for detecting oblique cracks on the rail surface are designed. Ultimately, a complete AC electromagnetic field-based system for detecting oblique cracks on the rail surface is established. The system is used to detect oblique cracks with different lengths and depths on the rail surface

Fig. 17 Results of crack array probes of different lengths. **a** Crack length of 6 mm. **b** Crack length of 12 mm. **c** Crack length of 18 mm. **d** Crack length of 24 mm. **e** Crack length of 30 mm

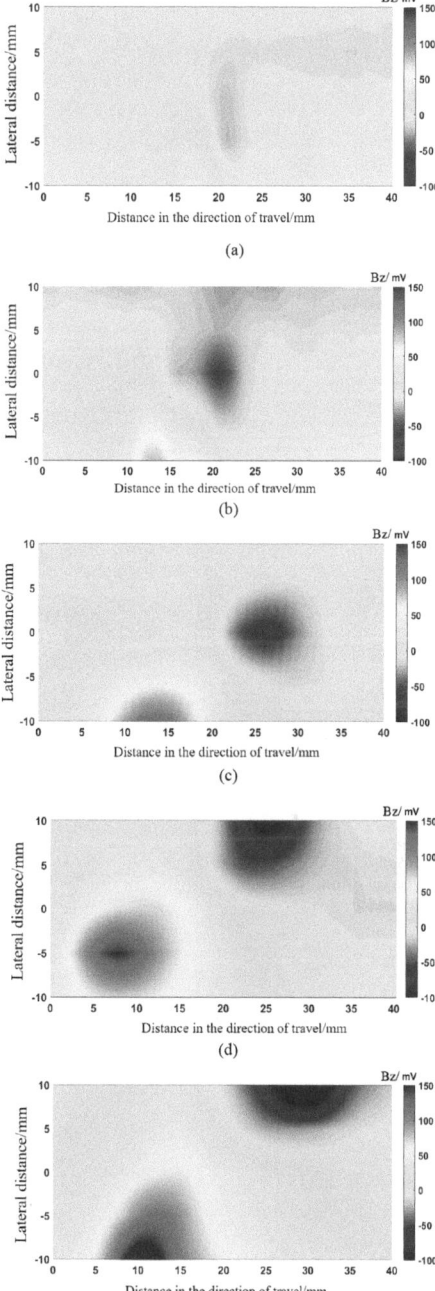

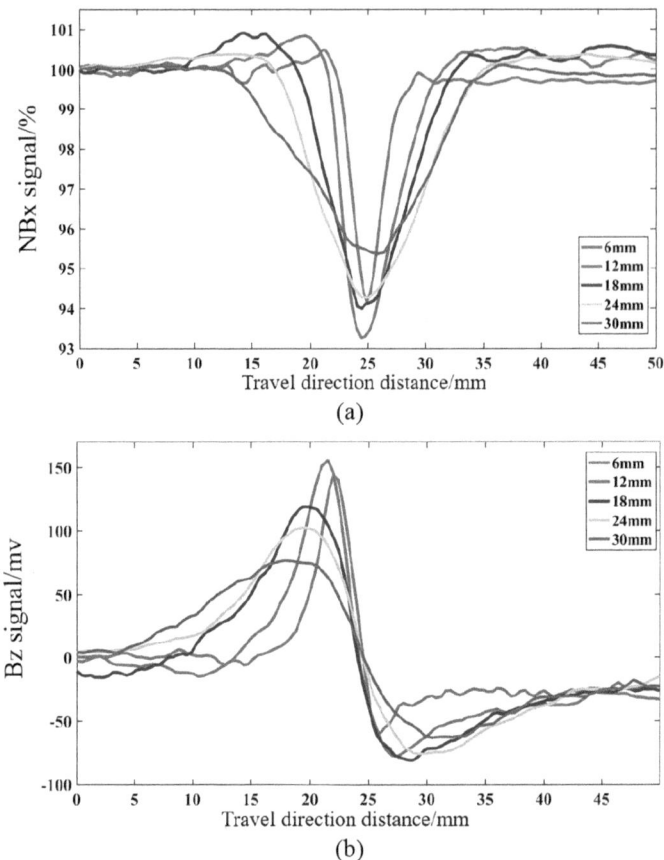

Fig. 18 Results of single probe for cracks of different lengths. **a** BX signal. **b** BX signal

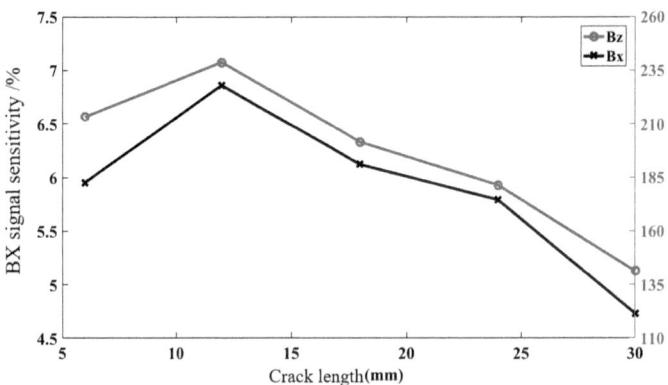

Fig. 19 The influence of crack length on characteristic signals

(1) The AC electromagnetic field detection method can be used to detect oblique cracks on the rail surface. When an oblique crack occurs, the induced current flows around the two endpoints of the crack, one in a clockwise direction and the other in a counterclockwise direction. The cross-sectional current flows around the bottom of the crack. As a result, the Bx signal at the center of the crack exhibits a trough, while the Bz signal exhibits peaks and troughs.

(2) The distortion amplitudes of the Bx and Bz signals vary depending on the scanning path. Since the induced current tends to concentrate at the endpoints of the crack, the distortion amplitudes of the Bx and Bz signals are minimal at the center of the crack. On the other hand, the distortion amplitudes of the Bx and Bz signals are maximal near the endpoints of the crack.

(3) The detection system can effectively detect oblique cracks with a depth of 1 mm on the rail surface and measure the length of the crack. For C-scan images, the distortion amplitude of a crack with a depth of 1 mm is 36 mV, while that of a crack with a depth of 7 mm is 209 mV. The measurement error of the crack length is less than 2 mm. The deeper the crack, the larger the distortion amplitude of the Bx and Bz signals at the center of the crack. Additionally, the distortion amplitudes of the Bx and Bz signals at the center of the crack increase and then decrease as the length of the crack increases.

References

1. Jia X (2020) Research on contact fatigue damage of rail based on wheel-rail transient rolling contact model. China Academy of Railway Sciences, Beijing
2. Xing L (2008) Research on defect characteristics and classification of higher speed rails. China Academy of Railway Sciences, Beijing
3. Zerbst U, Lundén R, Edel KO et al (2009) Introduction to the damage tolerance behaviour of railway rails: a review. Eng Fract Mech 76(17):2563–2601
4. Zhou Q, Zhang J, Guo Z et al (2010) Research on the rail damages and the preventive countermeasures in heavy haul railways. China Railway Sci 31(1):27–31
5. Shen J, Zhou L, Rowshandel H et al (2020) Prediction of RCF clustered cracks dimensions using an ACFM sensor and influence of crack length and vertical angle. Nondestruct Test Eval 35(1):1–18
6. Tian G, Gao B, Gao Y et al (2016) Review of railway rail defect non-destructive testing and monitoring. Chin J Sci Instr 37(8):1763–1780
7. Li G, Huang P, Chen P et al (2011) Application of eddy current testing in the quantitative evaluation of the rail cracks. J Zhejiang Univ 45(11):2038–2042, 2049
8. Lao J, Lu C (2013) Application of portable ultrasonic phased array instrument for rail welds ultrasonic inspection. Adv Mater Res 717(1):384–389
9. Gao Y, Wang P, Tian G et al (2013) Velocity effect analysis of dynamic magnetization in high speed detection for rail crack using MFL method. Nondestruct Test 35(10):53–58
10. Antipov A, Markov A (2014) Evaluation of transverse cracks detection depth in MFL rail NDT. Russ J Nondestr Test 50(8):481–490
11. Yu H, Li Q, Tan Y et al (2019) A coarse-to-fine model for rail surface defect detection. IEEE Trans Instrum Meas 68(3):656–666
12. Li T, Shi Y, Chen F et al (2021) Quantitative method of rail flaws based on ultrasonic phased array and total focusing DAC mappings. J Mech Eng 57(18):32–41

13. Li J (2017) Research on rail flaw detection based on time-frequency and space feature. South China University of Technology, Guangzhou
14. Ma W, Liu D, Zhao W (2010) A portable eddy current flaw detection for railway testing. Mach Des Manuf 2:88–90
15. Li N (2021) Research on rail surface defect detection system based on machine vision. Beijing Jiaotong University, Beijing
16. Zhao J, Li W, Ge J et al (2020) Coiled tubing wall thickness evaluation system using pulsed alternating current field measurement technique. IEEE Sens J 20(18):10495–10501
17. Shen J, Zhou L, Rowshandel H et al (2015) Determining the propagation angle for non-vertical surface-breaking cracks and its effect on crack sizing using an ACFM sensor. Measur Sci Technol 26(11):1–11
18. Yuan X, Li W, Yin X et al (2020) Visual reconstruction of irregular crack in austenitic stainless steel based on ACFM technique. J Mech Eng 56(10):27–33
19. Nicholson G, Davis C (2012) Modelling of the response of an ACFM sensor to rail and rail wheel RCF cracks. NDT&E Int 46(1):107–114
20. Rowshandel H, Nicholson GL, Shen JL et al (2018) Characterisation of clustered cracks using an ACFM sensor and application of an artificial neural network. NDT&E Int 98:80–88
21. Shen J, Liu M, Dong C et al (2022) Analysis on asymmetrical RCF cracks characterisation using an ACFM sensor and the influence of the rail head profile. Measurement 194:111008
22. Chacon Muñoz JM, García Márquez FP, Papaelias M (2013) Railroad inspection based on ACFM employing a non-uniform B-spline approach. Mech Syst Signal Process 40(2):605–617
23. Ge J, Yan C, Hu B et al (2021) High-speed detection of surface crack on rail using alternating current field measurement technique. J Mech Eng 57(18):66–74

Design and Testing of High-Resolution Probe Arrays Using Alternating Current Field Measurement Technique

Alternating Current Field Measurement (ACFM) technology is a new type of electromagnetic non-destructive testing technology, which can detect surface and near-surface defects of metal parts [1, 2]. This technique is based on the principle of electromagnetic induction. When sinusoidal excitation current is loaded into the excitation coil, a local uniform current field will be induced on the specimen surface. When the uniform current encounters a defect, the current will generate a disturbance, resulting in a disturbance of the magnetic field around the defect. The detection and evaluation of the defect can be realized by picking up the spatially-distorted magnetic field signal [3–5]. Due to the characteristics of non-contact measurement, small removal effect and high quantitative accuracy, ACFM technology has been widely used in petrochemical, railway transportation, nuclear power and other fields [6–8].

From the current research, the visualization methods of the existing ACFM technology and devices are sometimes based on magnetic field signal graphics such as scan map, butterfly map and isoline color map [9]. Compared with the previous two, isoline color map of magnetic field signal can not only determine the existence of defects, but also invert the appearance of defects. At present, there are two methods to obtain the isoline color map of magnetic field signals. One is to use a single probe to conduct two-dimensional scanning of the detected object. For example, Yuan [10] uses a three-axis bench to drive a single probe to conduct raster scanning of the irregular cracks on the surface of austenitic stainless steel to obtain the vertical magnetic field Bz image, and uses the gradient field visualization reconstruction method of Bz image to reconstruct the irregular cracks. However, the efficiency of this detection method is low, if you want to get a higher resolution magnetic field image, you need to conduct multiple scans. The other is to scan through array detection probe. Hu et al. [11] used LKACFM detector and 8 array detection probe to detect aluminum inner liner of high-pressure hydrogen cylinder, and obtained Bz images of round holes and cracks at different angles. Wu [9] designed 7 array detection probes to obtain three-dimensional magnetic field signals around rectangular slots and cylindrical corrosion pits, and used gradient fusion algorithm to invert surface profiles. Li

© The Author(s) 2025
W. Li et al., *Alternating Current Field Measurement Technique for Detection and Measurement of Cracks in Structures*, https://doi.org/10.1007/978-981-97-7255-1_5

[12] designed an orthogonal U-shaped array probe to detect aluminum plate specimens with surface corrosion defects, and proposed a defect surface shape inversion algorithm. Topp et al. [13] used the 8-array detection probe and detection system of TSC company to detect the railway, obtained the Bz scan image of the crack, and determined the Angle and length of the crack. At present, scholars and companies at home and abroad have designed relatively mature AC electromagnetic field array probes, but the sensor spacing of the existing probes is mostly 5 mm or above [14–17], resulting in low spatial resolution of the measured magnetic field image, which can only be used to determine the existence of defects and cannot accurately obtain the morphology of defects.

To solve this problem, an ACFM probe based on high resolution TMR sensor array is proposed in this paper, and the testing system is designed and built. The finite element software Comsol is used to establish an ACFM simulation model, analyze the electromagnetic field distortion rule around different types of defects, explore the relationship between the defect surface profile and magnetic field images, design a high-resolution ACFM array detection probe and test system, and carry out defect detection tests.

1 Finite Element Simulation

1.1 Model Building

The ACFM simulation model established by finite element software Comsol is shown in Fig. 1. The main components include: air region, excitation coil, U-shaped magnetic core and test piece to be tested. The entire model is surrounded by an air domain. The excitation coil is wound on the U-shaped magnetic core, the number of turns is 500, and the excitation signal uses a sinusoidal signal with a frequency of 1000 Hz and a current size of 150 mA. The sample is aluminum plate, located directly below the magnetic core, with a relative permeability of 1 and an electrical conductivity of 3.774×10^7 S/m. The extracted induced current on the specimen surface is shown in Fig. 2. The excitation coil generates a local uniform current field on the specimen surface, which meets the requirements of ACFM.

1.2 Defect Disturbance Law

In order to explore the disturbance law of induced electromagnetic field on the specimen surface which is caused by different types of defects, three defects are set on the specimen surface aiming at common cracks and area defects.

The cracks are 20 mm × 0.5 mm × 4 mm (length × width × depth), the spherical corrosion pit Φ20 mm × 4 mm (diameter × depth) and the square groove 20 mm ×

Fig. 1 Finite element
simulation model of ACFM

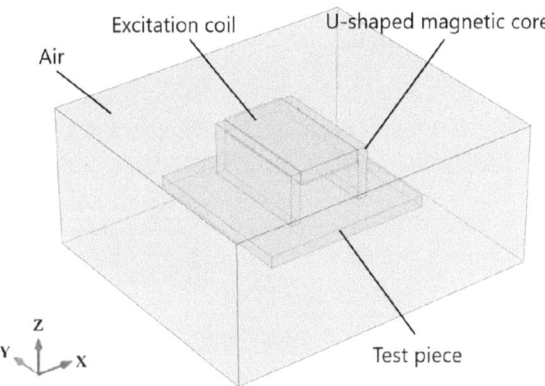

Fig. 2 Surface current
density

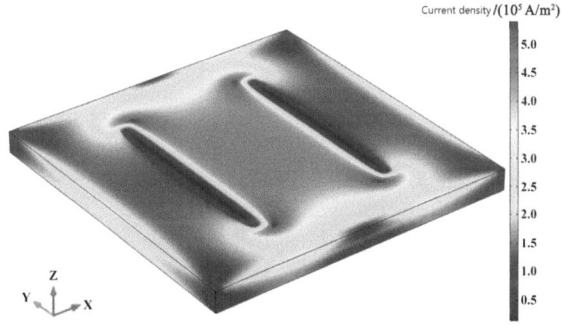

4 mm (side length × depth). The induced currents near the three defects are extracted
as shown in Figs. 3, 4 and 5. It can be seen that the uniform current field is disturbed
due to the existence of the defects. As shown in Figs. 3a and 5a, surface current
accumulates at the end points of the crack and the square groove, resulting in a
maximum current density at the end points. Similarly, although the surface profile
of the spherical corrosion pit is circular with no obvious tip, the surface current still
bypasses both sides of the arc, as shown in Fig. 4a, so the current density also presents
a maximum value on both sides of the arc. At the same time, due to the discontinuity
of the material, the current flows from the edge to the depth of the defect, as shown
in Figs. 3b, 4b and 5b, so the surface current density troughs in the defect region.

The distorted current causes the spatial magnetic field disturbance above the
defects. The simulation software adopts parametric scanning mode to simulate the
detection process of the actual probe, and extract the magnetic fields signal Bx and
Bz above 1 mm on the three defects. The scanning results are shown in Figs. 6, 7 and
8. Bx signal shows obvious troughs and Bz signal shows peaks and troughs, which
conform to the principle of ACFM. Moreover, it can be seen from Figs. 7a and 8a
that the shape of the trough in the Bx magnetic field image is roughly round and
square. As shown in Figs. 6b and 8b, the peak and valley distortion position in the
Bz image is the same as the current accumulation position at the defect endpoint.

Fig. 3 Disturbance law of current around crack

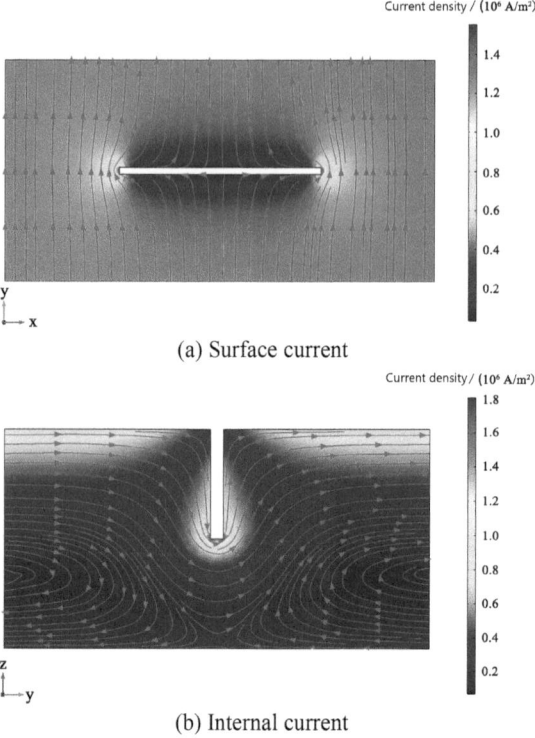

(a) Surface current

(b) Internal current

The simulation results show that the induced current accumulates at the end of the defect, and the accumulation current causes the distortion of the magnetic field Bz in the Z direction. So the magnetic field image of Bz contains the position information of the defective endpoint. At the same time, the current deflects from the edge of the defect to the bottom, causing a disturbance in the X direction magnetic field Bx, and the Bx magnetic field image can reflect the edge profile information of the defect.

2   Probe Design

2.1   Sensor Array Design

According to the simulation results, the amplitudes of the disturbed magnetic field Bx and Bz above the defect are in the order of 10–3 T and 10–4 T respectively, and the magnetic fields are relatively weak. Currently, commonly used magnetic field sensors include coil, Hall, AMR, GMR and TMR sensors. Compared with other types of magnetic field sensors, TMR sensors are more suitable for weak magnetic field detection because of their large linear range, high precision and high sensitivity

Fig. 4 Disturbance law of current around corrosion pit

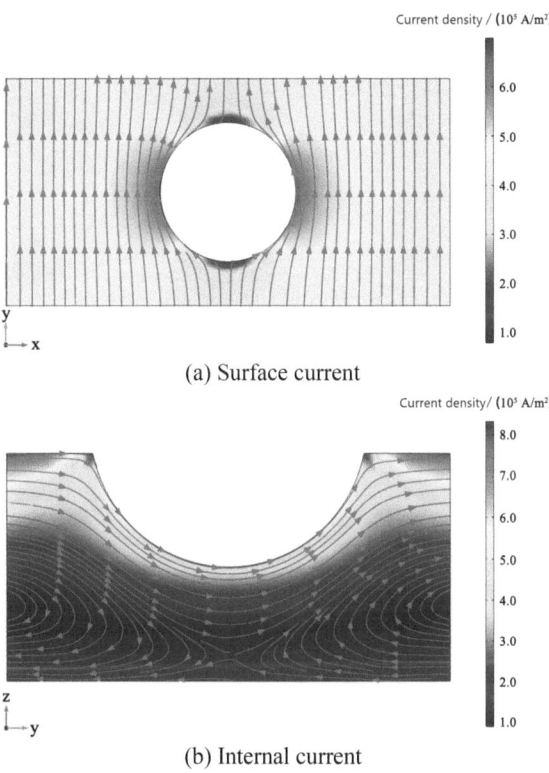

(a) Surface current

(b) Internal current

[18, 19]. In addition, the small package size of TMR sensors can be made very compact and dense, thereby improving the resolution in a small range [20], so TMR sensors are very suitable for array applications with high spatial resolution, weak magnetic field signal and high sensitivity.

Therefore, the TMR sensor is selected as the detection sensor in this paper. In order to ensure that the measured magnetic field image has a high spatial resolution while taking into account the manufacturing and processing cost, the sensor spacing is set to 1 mm. In order to ensure that the designed array probe has a certain detection range, the sensor array is composed of 64 TMR sensors, and the linear arrangement is adopted, so the effective detection range of the sensor array is 64 mm. The designed high-resolution sensor array is shown in Fig. 9. The black area at the bottom is the location of the sensor array, and the red arrow represents the direction of the sensor's sensitive axis. Each sensor and three resistors form a Wheatstone bridge to measure the magnetic field by measuring the change of resistance value. At the same time, the entire circuit board is supplied with power, and three through-holes are arranged in the middle of the circuit board to facilitate the sensor to be fixed when making the probe.

Fig. 5 Disturbance law of current around square groove

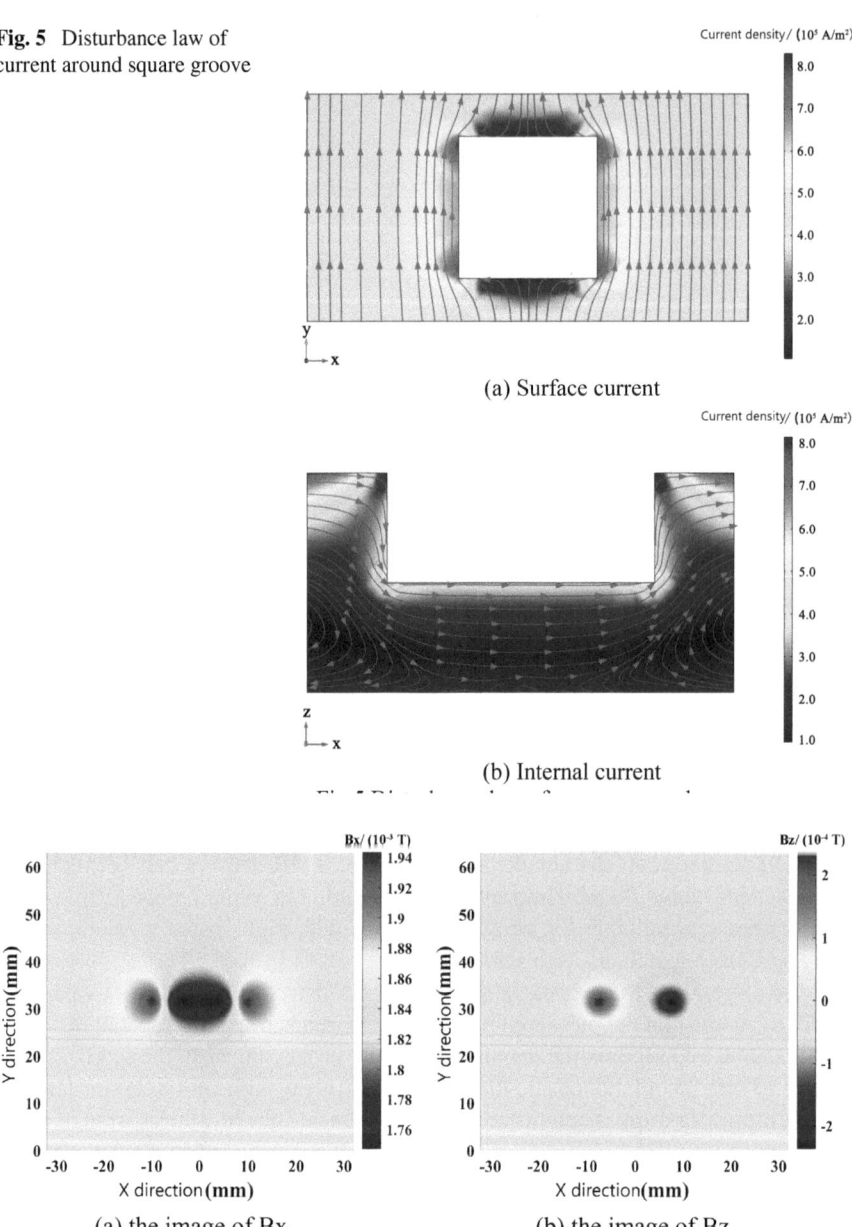

(a) Surface current

(b) Internal current

(a) the image of Bx

(b) the image of Bz

Fig. 6 Magnetic field image above the crack

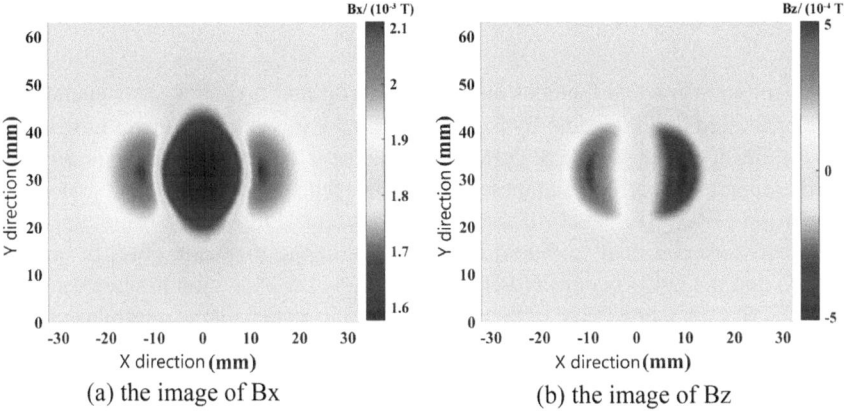

(a) the image of Bx   (b) the image of Bz

Fig. 7 Magnetic field image above the corrosion pit

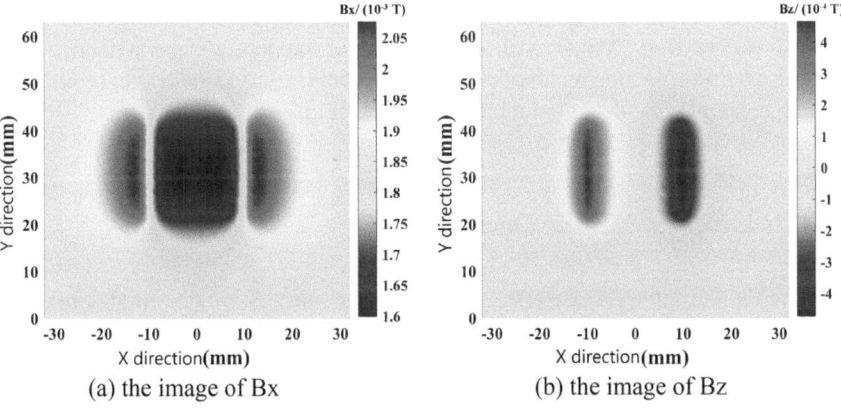

(a) the image of Bx   (b) the image of Bz

Fig. 8 Magnetic field image above the square groove

Fig. 9 TMR sensor array

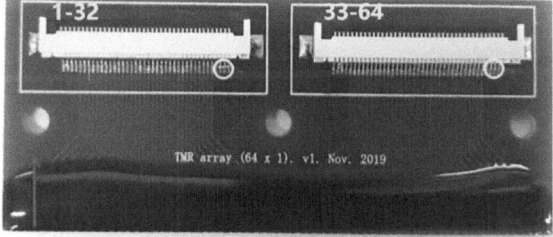

2.2 Excitation Module Design

The excitation module includes an excitation coil and a U-shaped magnetic core. The excitation coil is wound by a enamel-coated wire with a diameter of 0.15 mm on the beam of the magnetic core, with a number of turns of 500. The material of the magnetic core is manganese zinc ferrite, which can effectively enhance the excitation effect. The width of the uniform current field on the specimen surface is directly proportional to the width of the U-shaped magnetic core. In order to ensure that the uniform current field stimulated by the excitation module can cover the effective detection range of the sensor array, this paper adopts multiple magnetic cores placed side by side to increase the width of the uniform current field. The width of the selected single magnetic core is 16.1 mm, which is simulated by the simulation software Comsol. Finally, 5 magnetic cores are selected to form an exciting magnetic core array, and the induced current under the magnetic core array is extracted, as shown in Fig. 10. The red area in Fig. 10a is the uniform current field, and Fig. 10b is the induced current density at the location of the sensor array (directly below the magnetic core array, $X = 0$ mm). 90% or more of the peak current is defined as the uniform current, and the width of the uniform current area is obtained to be 66.3 mm, which can cover the detection range of the sensor array.

2.3 Multiplex Module Design

To process and collect the signals picked up by the sensor array, a multi-channel signal processing circuit and an AD acquisition channel are needed. However, too many processing circuits will cause interference between signals, and too many acquisition channels will also cause the sampling rate to slow down, which will affect the detection effect. Therefore, multiplex method is used to pre-process the multi-channel signals output by sensor array. According to the time division transmission mode and frequency division transmission mode, the multiplexing technology is divided into time division multiplexing and frequency division multiplexing. In this paper, time division multiplexing technology is adopted [21], and ADG1607 multiplexing chip is selected to achieve multiplexing. As shown in Fig. 11, the chip has a single-chip iCMOS® analog multiplexer with 8 differential channels. One of the eight differential inputs can be switched to a common differential output at an address determined by the 3-bit binary address lines (A0, A1, and A2), and the chip conversion time is 143 ns. The delay time during switching is negligible, and there is no significant difference between the output and input signals, so there is almost no impact on the final detection result.

A single ADG1607 chip can switch 16 differential inputs to 2 common differential outputs, as shown in Fig. 11b. In order to realize the multiplexing of the output signals of the sensor array, four multiplexing chips are set in the middle of the circuit board to switch the 64 sensor output signals to 8 outputs, as shown in Fig. 12a. The design

Fig. 10 Simulation analysis of five U-shaped cores

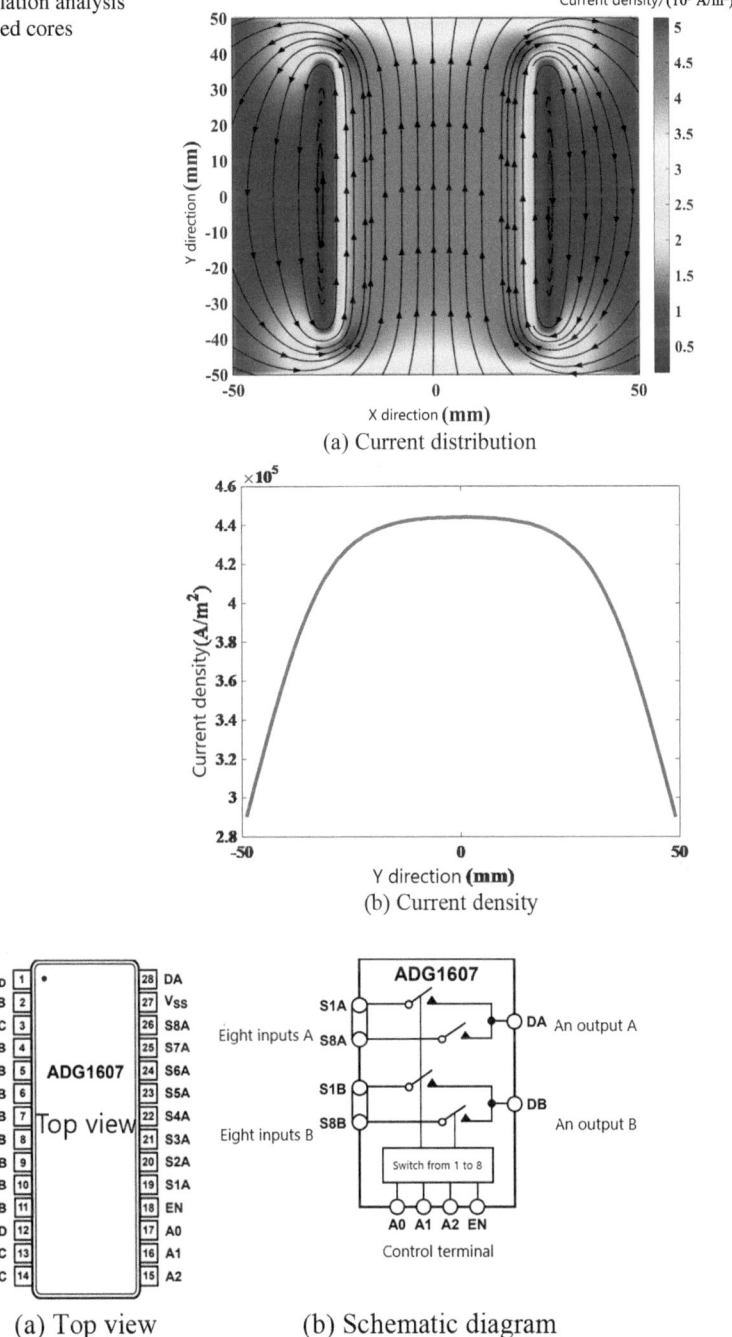

(a) Current distribution

(b) Current density

(a) Top view   (b) Schematic diagram

Fig. 11 ADG1607 multiplexer chip

Fig. 12 Signal multiplexing module

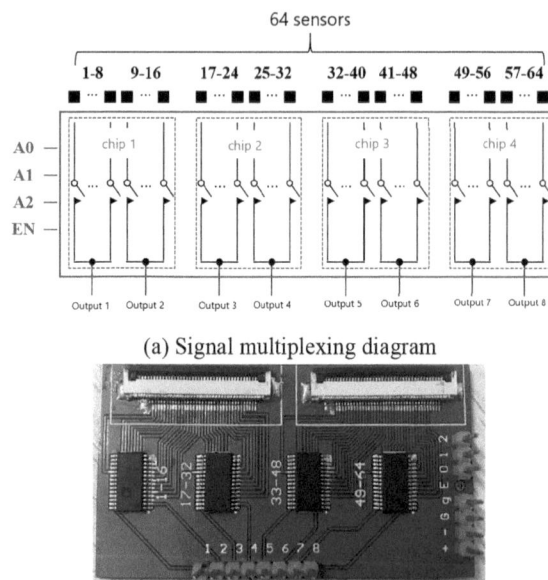

(a) Signal multiplexing diagram

(b) Multiplex circuit

of the multiplex module is shown in Fig. 12b. The upper two 40-pin connectors are connected to the sensor array circuit board through the FPC SoftBank cable, and the right side is the power supply interface and the analog switch control port. At the same time, the EN is always set to high level, and the switch is switched by the binary digital address line control chip output by the digital output channel of the acquisition card, and the 64 signals are finally converted to 8 signals. This process simplifies the hardware processing circuit.

2.4   Design of Probe Structure

The array detection probe is mainly composed of high-resolution TMR sensor array, U-core array, excitation coil, multiplexing module, top cover, housing and Remo socket. The structure of the detection probe is shown in Fig. 13. A groove is arranged at the bottom of the interior of the housing for fixing the sensor circuit board, which can be used to detect Bx signals. The sensor circuit board is placed vertically through the fixing block, which can be used to detect Bz signals. The magnetic core array is located directly above the sensor array, and a card slot is arranged inside the housing to ensure that the magnetic core array will not move during detection. The top cover is provided with a hole according to the size of the Remo socket to fix the Remo socket, to achieve stable internal signal transmission and power supply for the probe. The top cover is fixed to the housing by countersunk screws.

Fig. 13 Detection probe array

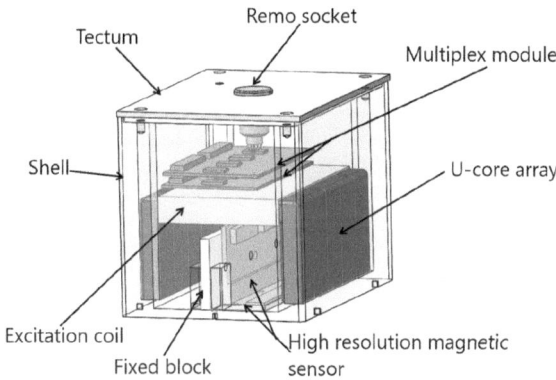

Tectum — Remo socket — Multiplex module — Shell — U-core array — Excitation coil — Fixed block — High resolution magnetic sensor

3   Experimental Study

3.1   Experimental System

The whole test system, as shown in Fig. 14, mainly consists of array detection probe, detection instrument (hardware chassis and industrial PC integration) and defect detection software. The power module is powered by the PC battery and supplies power to the entire detection system through the voltage regulator and voltage conversion modules. The excitation module of the detector loads the sinusoidal excitation signal with frequency of 1 kHz and amplitude of 10 V onto the excitation coil of the array probe, and the excitation coil induces a uniform electric field on the surface of the specimen. When the probe passes through the defect, the uniform electric field on the surface of the specimen is disturbed due to the difference in the conductivity between the air and the specimen, resulting in the disturbance of the magnetic field above the specimen. The sensor array in the detection probe picks up the distorted spatial magnetic field signal, preprocesses the signal through the multiplexing module, and then passes the signal conditioning (amplification and filtering), and finally transmits the signal to the signal acquisition module for AD conversion and transmission to the PC. The software on the PC is developed based on Labview to collect, process, store and display the magnetic field signal around the defect.

3.2   Experimental Test

The developed test system is used to detect the artificial defects of the prefabrication. The test specimen was aluminum plate, and the defect type was consistent with the simulation, as shown in Fig. 15. The crack size is 12 mm × 0.5 mm × 3 mm (length × width × depth), and the Angle between the crack and the probe scanning direction is 0°, 30° and 60°, respectively. The depth of the three pits is 3 mm, and the diameters

Fig. 14 Test system

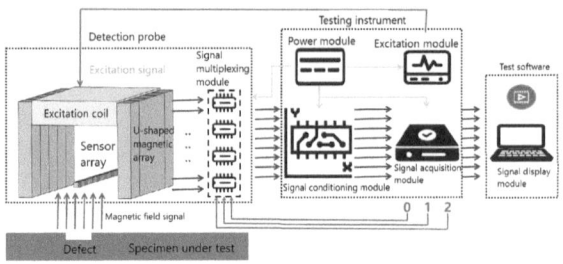

(a) Schematic diagram of system design

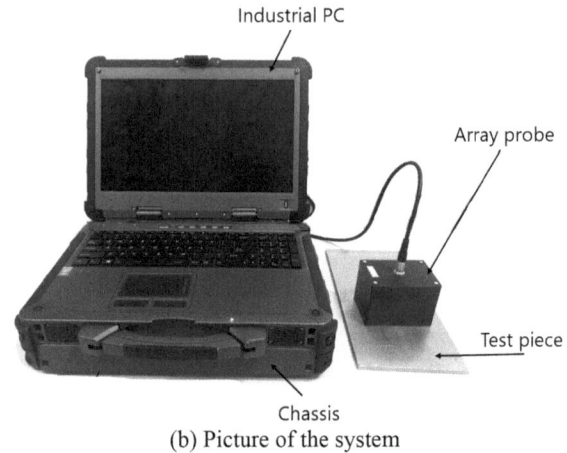

(b) Picture of the system

are 10 mm, 14 mm and 18 mm, respectively. The depth of the three square slots is 3 mm and the side lengths are 6 mm, 10 mm and 15 mm respectively. The probe was placed on the surface of the specimen, and the three types of defects were scanned successively with a scanning step length of 0.5 mm. After the scan is completed, the magnetic field images of Bx and Bz are obtained, as shown in Figs. 16, 17 and 18. The X-direction magnetic field Bx of the three defects showed troughs, and the Z-direction magnetic field Bz showed peaks and troughs, which was consistent with the simulation results. It can be seen that the ACFM array probe and system designed in this paper can obtain the distorted magnetic field signal around the defect.

3.3 Defect Inversion Imaging

According to the simulation results, it can be seen that the Bz image can reflect the endpoint position of the defect, and the Bx image has the edge information of the defect, so the surface profile of the defect can be retrieved through the images of the magnetic field signals Bx and Bz. The crack width is very narrow and there is no obvious surface profile, so the end point of the crack can be determined using the Bz

Fig. 15 Schematic diagram of atificial defects

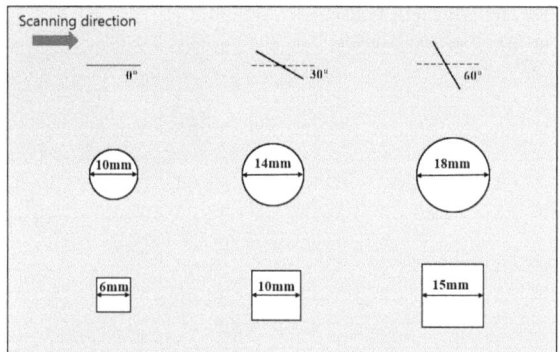

Fig. 16 Magnetic field image of cracks

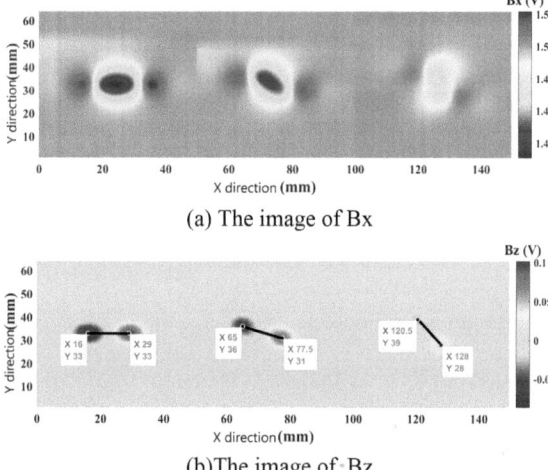

(a) The image of Bx

(b)The image of Bz

Fig. 17 Magnetic field image of corrosion pits

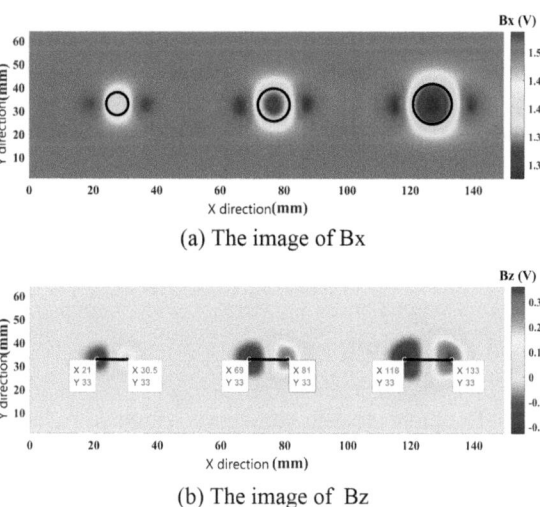

(a) The image of Bx

(b) The image of Bz

Fig. 18 Magnetic field
image of square grooves

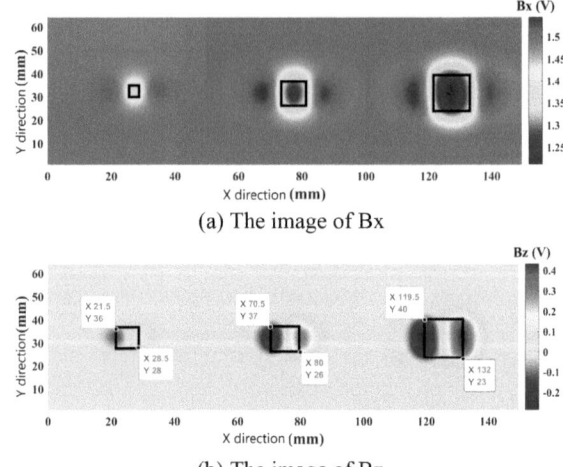

(a) The image of Bx

(b) The image of Bz

image, and the length and angle of the crack can be determined by connecting the
end point, as shown in Fig. 16b. The evaluation results of the three cracks are 13 mm
× 0°, 13.4 mm × 21.8°, 13.3 mm × 55.7°, respectively. The maximum evaluation
error of the length was 1.4 mm, and the maximum evaluation error of the Angle is
8.2°.

For area-type defects of the corrosion pit type, the surface profile is relatively
obvious but there is no obvious tip. The contour of the defect can be retrieved
according to the Bx image, as shown in Fig. 17a. It can be seen that the Bx image is
consistent with the surface profile of the corrosion pit, and the distance between the
peak and valley in the Bz image is the diameter of the surface profile, as shown
in Fig. 17b. The diameter quantization results are 9.5 mm, 12 mm and 15 mm
respectively, and the maximum quantization error is 3 mm.

For square slot area defects with relatively obvious tip and surface profile, surface
topography can be inverted by Bx images. As shown in Fig. 18a, the Bx image of a
square slot basically presents a square. The peak and valley distortion positions in
the Bz image are connected, which can simultaneously invert the surface topography
and quantify the side length of the square slot. As shown in Fig. 18b, the side length
quantization results are 7 mm × 8 mm, 9.5 mm × 9 mm, 12.5 mm × 17 mm,
respectively, and the maximum quantization error is 2.5 mm.

4 Conclusion

(1) The finite element simulation results of AC electromagnetic field show that the
    induced current accumulates at the end of the defect, resulting in the distortion
    of the magnetic field Bz in the Z direction, and the Bz image can reflect the end

information of the defect. The induced current bypasses the edge of the defect to the bottom, causing the disturbance of the Bx signal of the magnetic field in the X direction, and the Bx image can reflect the contour information of the defect.

(2) The ACFM array detection probe based on high-resolution TMR sensor array designed in this paper can obtain horizontal (X) and vertical (Z) magnetic field images of different types of defects, and the magnetic field images have a spatial resolution of 1 mm.

(3) The obtained high-resolution magnetic field image can accurately invert the surface profile of the defect. Bz image can reconstruct the length and Angle of the crack, and Bx image can invert the surface profile of the area defect. The maximum evaluation error of crack length was 1.4 mm. The maximum evaluation error of Angle is 8.2°. The maximum quantization error of corrosion pit diameter is 3 mm. The maximum quantization error of square slot side length is 2.5 mm. These results show that the system has high evaluation accuracy.

References

1. Salemi AH, Sadeghi SHM, Moini R (2004) Thin-skin analysis technique for interaction of arbitrary-shape inducer field with long cracks in ferromagnetic metals. NDT&E Int 39:471–479
2. Wei FAN (2014) Design and test of the AC field detection system. Nanchang Hangkong University, Nanchang
3. Lewis AM, Michael DH, Lugg MC et al (1998) Thin-skin electromagnetic fields around surface-breaking cracks in metals. J Appl Phys 64(08):3777–3784
4. Yuan XA, Li W, Chen GM et al (2018) Two-step interpolation algorithm for measurement of longitudinal cracks on pipe strings using circumferential current field testing system. IEEE Trans Industr Inf 14:394–402
5. Zhao J, Li W, Yuan X et al (2021) Lift-off suppression algorithm for dual-frequency alternating current field measurement. Nondestruct Test 43(04):5–9
6. Li W, Yuan X, Chen G et al (2015) Research on in-service detection for axial cracks on drill pipe using the feed-through alternating current field measurement. J Mech Eng 51(12):8–15
7. Shen JL, Zhou L, Rowshandel H et al (2015) Determining the propagation angle for non-vertical surface-breaking cracks and its effect on crack sizing using an ACFM sensor. Meas Sci Technol 26(11):1–11
8. Smith M, Laenen C (2007) Inspection of nuclear storage tanks using remotely deployed ACFMT. Insight 49(01):17–20
9. Wu Y (2017) Research on three-dimensional visualization of defects based on ACFM. China University of Petroleum (East China), Qingdao
10. Yuan X, Li W, Yin X et al (2020) Visual reconstruction of irregular crack in austenitic stainless steel based on ACFM technique. J Mech Eng 56(10):27–33
11. Hu D, Li D, Chen Z et al (2021) Detection technology of ac electromagnetic field in aluminum liner of high-pressure hydrogen cylinder. Low Temp Spec 39(02):42–45
12. Li W (2007) Research on ACFM based defect intelligent recognition and visualization technique. China University of Petroleum, Dongying
13. Topp D, Smith M (2005) Application of the ACFM inspection method to rail and rail vehicles. Insight 47(06):354–357
14. Li W, Yuan X, Chen G et al (2017) Research on the detection of surface cracks on drilling riser using the chain alternating current field measurement probe array. J Mech Eng 53(08):8–15

15. Zhao JM, Li W, Zhao JCH et al (2020) A novel ACFM probe with flexible sensor array for pipe cracks inspection. IEEE Access 8:26904–26910
16. Mirshekar-Syahkal D, Mostafavi RF (2002) 1-D probe array for ACFM inspection of large metal plates. IEEE Trans Instrum Meas 51(02):374–382
17. Topp D (2005) Recent developments and applications of the ACFM inspection method and ACSM stress measurement method. Non-Destruct Test Austr 42(5):141–147
18. Zhang N, Peng L, Wu Y et al (2020) ECT probe based on magnetic field imaging with a high resolution TMR sensor array. Chin J Sci Instrum 41(07):45–53
19. Gao J, Wang J, Qiu F (2017) Magnetic sensitive automobile detector based on TMR sensor. Chin J Sci Instrum 38(08):2039–2046
20. Tao Yu, Lv K, Peng L et al (2020) Design of probe for heat exchange tubes detection of steam of steam generator based on TMR sensor array. Instrum Techn Sensor 05:37–47
21. Liu H, Xiao B, Wang L (2007) Parameter detection of a certain missile control casing based on virtual instrument technology. Chin J Sci Instrum 28(04):257–259

Design and Experimental Study of Inner Uniform Electromagnetic Probe in Stainless Steel Pipe

Petroleum and petrochemical industry is a pillar industry of the national economy, which occupies a pivotal position in the national economy. In petroleum and petrochemical, stainless steel pipelines are widely used in pipeline transportation systems, including high-pressure furnace pipes, piping, petroleum cracking pipes, fluid delivery pipes, heat exchange pipes, etc. But due to the corrosive medium in the pipeline and the alternating cooling and heating environment, fatigue cracks are easy to occur on the inner wall of the pipeline [1–3], usually, fatigue cracks aggregate and grow in the longitudinal direction [4], which eventually leads to leakage and failure of the pipeline, so it is necessary to regularly detect the pipeline.

At present, conventional detection technologies mainly include magnetic particle, penetration, ultrasound, magnetic flux leakage and eddy current detection technologies [5]. Magnetic particle and penetration detection technology is currently the most reliable non-destructive detection technology, but magnetic particle and penetration detection need to be in direct contact with the tested material, the surface requirements are relatively high, and only crack detection can be realized, cracks cannot be quantified, and detection in the pipeline is difficult to achieve [6]; ultrasonic detection technology needs to add a coupler between the detection tool and the tested pipeline, if it is not thoroughly cleaned [7], it is difficult to apply a coupler on the inner wall of the pipeline; magnetic flux leakage detection technology can only detect ferromagnetic materials [8]; eddy current detection technology is easily affected by lifting and jitter.

Uniform field disturbance nondestructive testing methods, such as alternating current electromagnetic field detection technology, alternating current potential drop detection technology and circumferential current field detection, have good prospects for detecting and evaluating conductive materials [9–11]. At present, a series of studies have been carried out on the form of uniform electromagnetic field detection probes, Li Wei and others have established a U-shaped probe simulation model and built an experimental system [12]; Yuan Xin'an and others have used double U-shaped probes to detect cracks in any direction [13]; Ge Haojiu and others have used

© The Author(s) 2025
W. Li et al., *Alternating Current Field Measurement Technique for Detection and Measurement of Cracks in Structures*, https://doi.org/10.1007/978-981-97-7255-1_6

external penetrating probes to achieve the detection of cracks on the outer wall of the pipeline [14, 15].

However, the U-shaped probe is not suitable for pipe surface detection. Due to the skin effect of the external penetrating probe, it is difficult to realize the detection of the inner wall of the pipeline. In order to realize the detection of the inner wall of the pipeline, the internal detection probe is also required. However, the research on the detection probe in the uniform electromagnetic field pipeline is still blank.

Aiming at the shortage that the current uniform electromagnetic field detection probe is not suitable for in-pipe detection, this paper proposes a Hermholtz coil-type aluminum alloy in-pipe detection probe and builds an experimental system. Through the finite element software Comsol, establish a simulation model for in-pipe detection of stainless steel pipes, analyze the distribution and change law of electromagnetic fields in stainless steel pipes, extract characteristic signals, analyze the influence of crack size on characteristic signals, design a stainless steel pipe in-pipe detection probe, build an in-pipe detection experimental system, and detect cracks on the inner wall of the pipeline.

1 Finite Element Emulation

The detection probe in the uniform electromagnetic field pipeline adopts the Hermholtz coil structure. The Hermholtz coil is a pair of parallel and coaxial coils with the same current-carrying coil. When the current in the same direction is passed to the coil and the distance between the two coils is equal to the coil radius, the total magnetic field of the coil will be uniformly distributed near the center of the axis. Based on the characteristics of the Hermholtz coil, a detection model in the pipeline is established to explore the distribution and change law of the electromagnetic field in the pipeline [15].

1.1 Distribution Law of Electric Field in Pipeline

According to the structure of the Helmholtz coil, the finite element software COMSOL is used to establish a simulation model for in-pipe detection as shown in Fig. 1. The specific model parameters are shown in Tables 1 and 2.

Extracting the current density of the inner surface of the pipeline is shown in Fig. 2 from the results, it can be seen that the maximum value of the induced current is near the excitation coil, and a uniform electric field is induced in the middle of the two coils, which meets the requirements of uniform electromagnetic field detection.

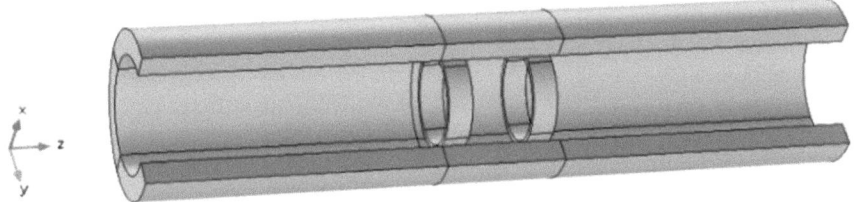

Fig. 1 Simulation model

Table 1 Sizes of model

Composition	Outer diameter (mm)	Inner diameter (mm)	Length (mm)
Pipeline	65	45	200
Excitation coil	40	36	10
Air	–	–	–

Table 2 Parameters of model

Wire diameter (mm)	Number of turns	Pipe material	Current size (A)	Magnetic permeability	Conductivity (S/m)	Excitation frequency (Hz)
0.8	200	Stainless steel	0.5	1	3.7e7	2000

Fig. 2 Surface current density

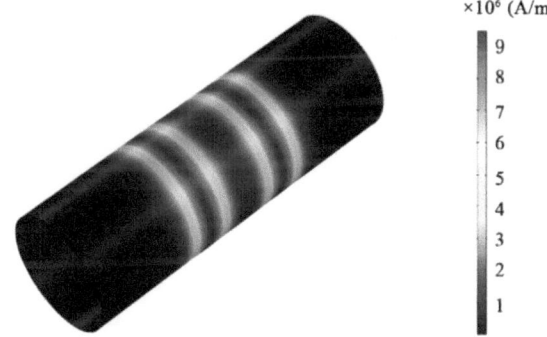

$\times 10^6$ (A/m²)

9
8
7
6
5
4
3
2
1

1.2 Crack Disturbance Law

In order to explore the disturbance mechanism of cracks, the electromagnetic field distribution law in the aluminum alloy pipeline when there are cracks on the inner surface of the emulation pipeline. The length of the axial crack on the inner wall of the pipeline is 20 mm, the width is 1 mm, and the depth is 5 mm.

Extracting the current density near the axial crack on the inner wall of the pipeline is shown in Fig. 3. The results show that when a uniform current field passes through the crack, as shown in Fig. 3a, the current on the surface will bypass the tip of the crack. As shown in Fig. 3b, the internal current will bypass from the bottom of the crack, which conforms to the principle of uniform electromagnetic field detection.

Using parametric scanning to simulate real probe detection, the origin of the coil is scanned from − 60 mm to 60 mm in the Z coordinate, and the step size is 1 mm. Considering the probe structure and the size of the sensor, the axial magnetic field signal and the radial magnetic field signal directly above the crack are extracted at 7 mm as shown in Fig. 4. It can be seen from the figure that when there is no crack, the signal is a fixed value, and the signal is 0. When passing through the defect, the signal appears two smaller troughs and a larger crest, and the signal appears a crest

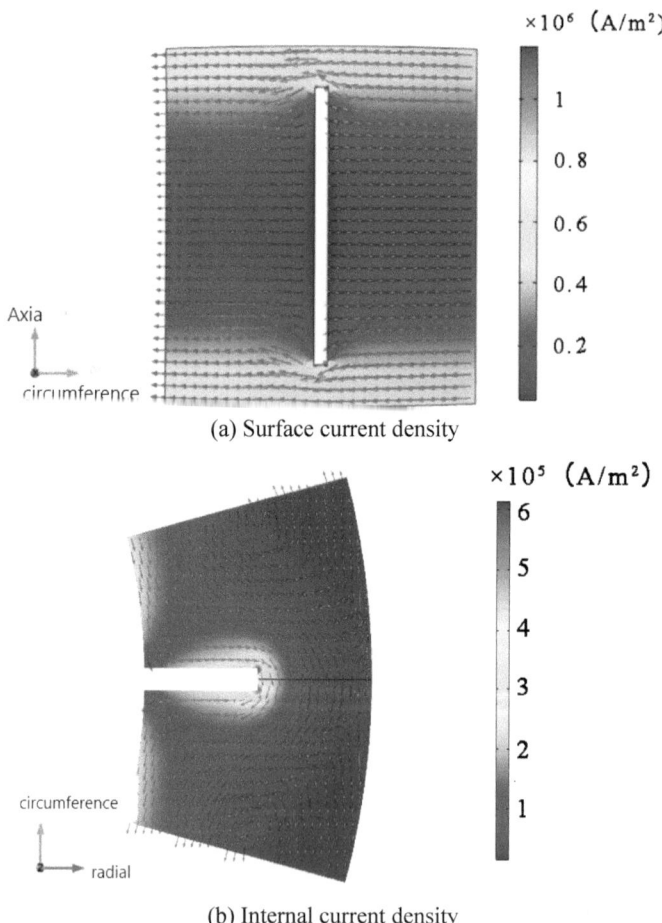

(a) Surface current density

(b) Internal current density

Fig. 3 Current dens

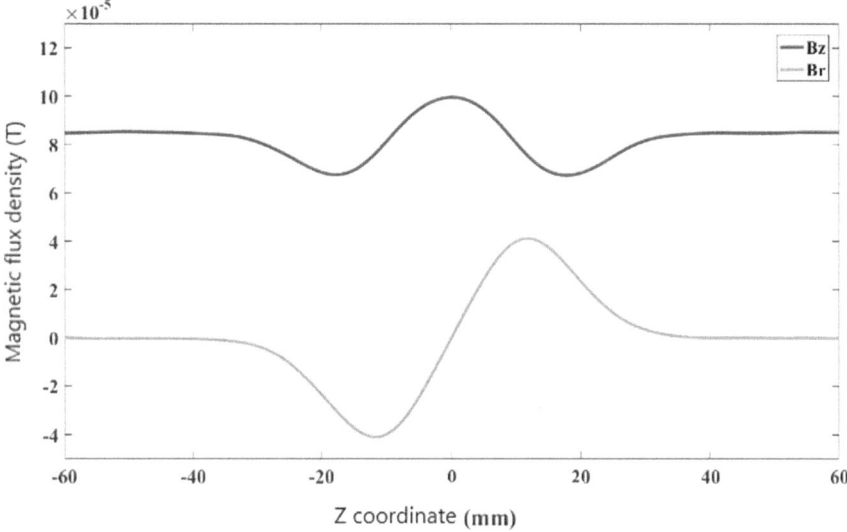

Fig. 4 Axial direction signal Bz and radial direction signal Br above crack

trough. Therefore, the axial magnetic field signal and the radial magnetic field signal are used as characteristic signals for crack detection.

1.3 Effect of Crack Size on Characteristic Signal

In order to explore the relationship between the extracted feature signal and the crack size, finite element emulation analysis is carried out on cracks of different depths and different lengths. First, keep the length of the crack (20 mm) unchanged, and emulation analysis is carried out on cracks of different depths (3, 4, 5, 6, 7 mm). The axial magnetic field signal at the 7 mm lift-off above the crack is extracted. The emulation result is shown in Fig. 5a. From the results, it can be seen that with the increase of the crack depth, the height of the crest also increases. It is defined as the abnormal variable of the axial magnetic field signal caused by the crack, and the background value is the difference value between the maximum value of the crest and the background value. The relationship between the obtained crack depth and the abnormal variable is shown in Fig. 5b. It can be seen that within a certain range of the crack depth, the abnormal and depth variable of the magnetic field maintain a relatively good linear relationship, indicating that the characteristic signal contains the crack depth information.

Keep the crack depth (5 mm) unchanged, emulation analysis of cracks of different lengths (10, 20, 30, 40, 50 mm), and extract the radial magnetic field signal at the 7 mm lift-off above the crack. The results of emulation are shown in Fig. 6a, b. As

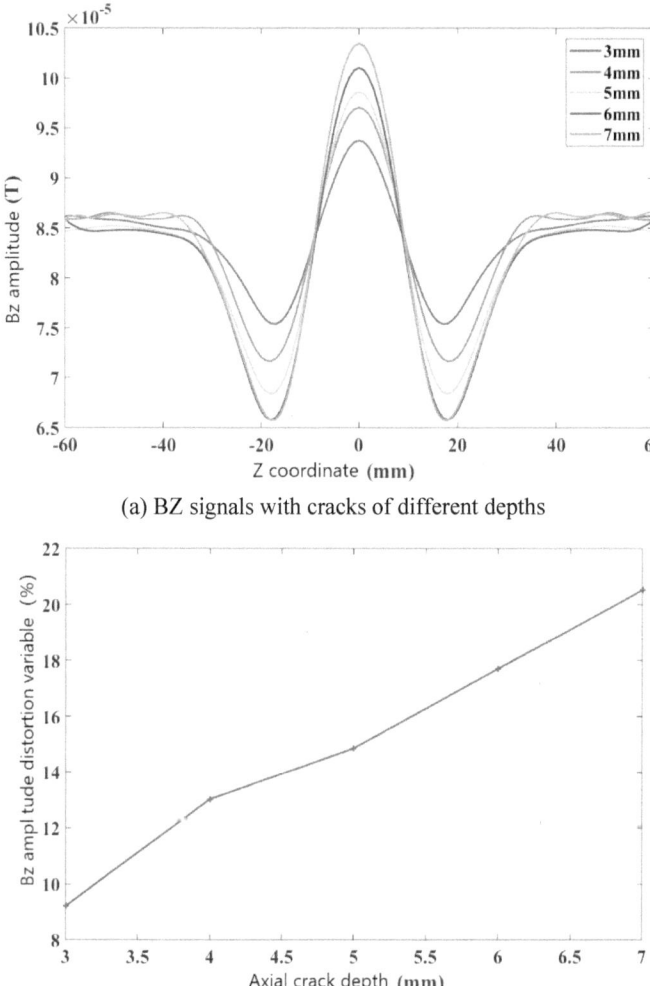

(a) BZ signals with cracks of different depths

(b) The amplitude of cracks at different depths is variable

Fig. 5 Relationship between BZ and the crack depth

shown, when the crack length is small, the crack length will affect the crest trough spacing and amplitude of the signal. The longer the crack, the greater the crest trough spacing and the distortion amplitude; when the crack is long, the crack length hardly affects the amplitude of the distortion, and the crest trough The spacing will increase with the increase of the crack length, so the characteristic signal contains the length information of the crack, especially for the detection effect of long cracks will be better.

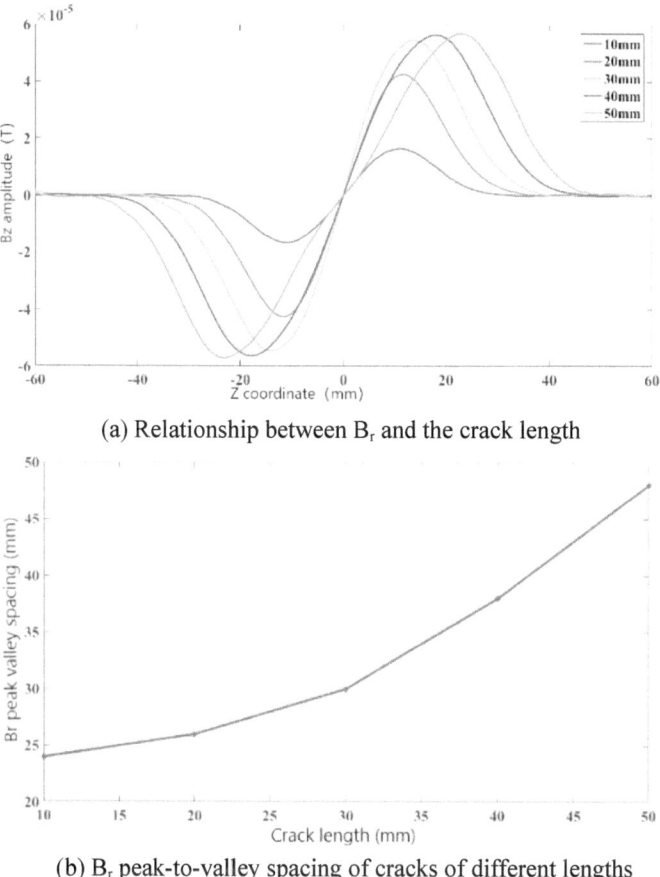

(a) Relationship between B_r and the crack length

(b) B_r peak-to-valley spacing of cracks of different lengths

Fig. 6 Relationship between B_r and the crack length

2 Detection System Construction and Experiment

2.1 Detection Probe Design

The structure of the detection probe is shown in Fig. 7, which mainly includes an excitation coil, a non-magnetic skeleton, a cable protection device, a connecting device and a detection sensor. A 0.8 mm enameled wire is wound on both sides of the non-magnetic skeleton for 200 turns to form a Hermholtz excitation coil, which is used to generate an excited magnetic field; the cable protection device is used to pass and protect the input and output of the excitation coil and the detection sensor; the connecting device is used to connect with the driving device in the pipeline to complete the automatic scanning of the pipeline.

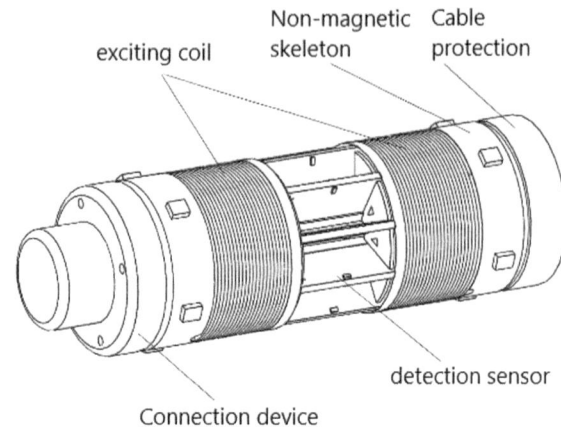

Ten detection sensors are evenly distributed in the circumferential direction of the middle position of the excitation coil, which can realize the comprehensive detection of cracks on the inner wall of the pipeline at one time. The installed circuit board is shown in Fig. 8, and the selected sensor is a small volume, large linear range, and high sensitivity TMR sensor. The sensor is installed near the edge of the center of the circuit board, and the TMR for detecting the axial characteristic signal is on the A side of the circuit board, as shown in Fig. 8a, and the TMR for detecting the radial characteristic signal is on the B side of the circuit board, as shown in Fig. 8b. The AD620 for amplifying the signal is installed on the circuit board, and the sum signal is amplified 100 times each. With the help of the co-ground capacitance, the ability of filtering, noise reduction and anti-interference is achieved [16, 17].

2.2 Detection System

The entire detection system as shown in Figs. 9 and 10, comprising a bench, pipeline under test, detection probe, hardware integrated chassis, PC. The signal generator in the hardware integrated chassis generates a sinusoidal AC signal of 2000 Hz, which is loaded on the excitation coil of the detection probe. The excitation coil induces a uniform electric field and magnetic field in the pipeline. The bench drives the detection probe to run at a uniform speed of 10 mm/s in the pipeline. When there is a defect in the pipeline, it will cause the electric field and magnetic field in the pipeline. Distortion, the circumferential direction of the intermediate position of the detection probe 10 evenly distribute the detection sensor picks up the magnetic field signal, and the acquisition card in the hardware integrated chassis is collected into the PC. The PC processes the signal based on the software identification program jointly developed by Matlab and Labview, and finally identifies the defect.

Fig. 8 TMR PCB

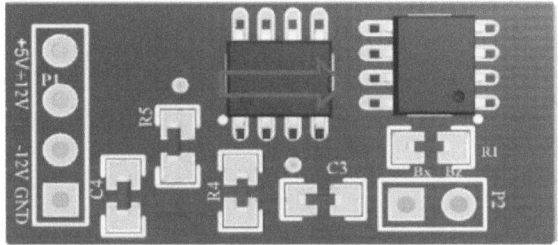

(a) A-side of TMR PCB

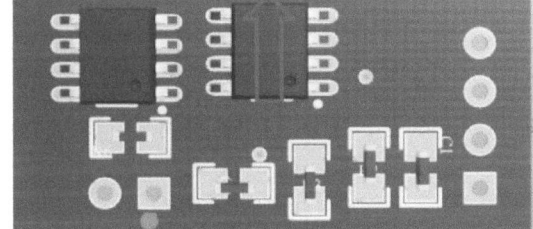

(b) B-side of TMR PCB

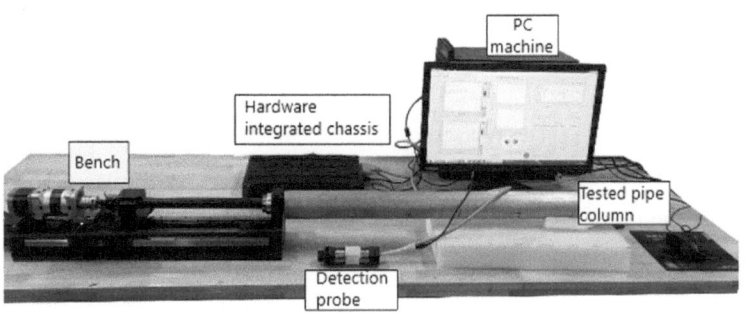

Fig. 9 Test system

2.3 Test Results

The cracks of different depths and the same length on the inner wall of the pipeline are detected, and the pipe size is shown in Fig. 11. The obtained sum signal is shown in Fig. 12. From the results, when there are cracks, the characteristic signal will crest, and the characteristic signal will crest and trough. For cracks of the same length and different depths, the deeper the crack, the greater the abnormal variable of the signal. Observe each characteristic signal, the abnormal variable of the No. 3 sensor is the largest, and it can be preliminarily determined that the defect exists near the No. 3 sensor. Extract the signal of the No. 3 sensor to obtain the relationship between the abnormal variable of the signal and the crack depth as shown in Fig. 14. The results

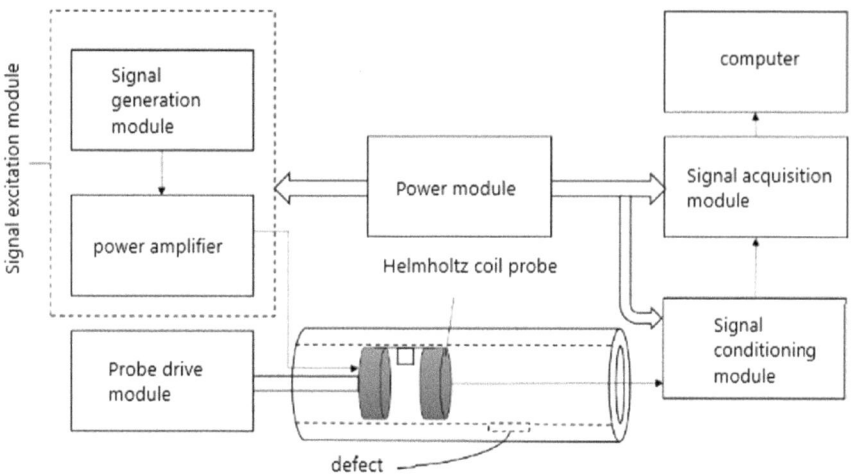

Fig. 10 Experiment system structure chart

crack width is 0.8mm
the unit is mm

Fig. 11 Pipes with different depth cracks

show that the depth of the crack has a good linear relationship with the abnormal variable of the signal, which is consistent with the emulation result (Fig. 13).

For cracks of different lengths and the same depth on the inner wall of the pipeline, the size of the pipeline is shown in Fig. 14. The obtained sum signal is shown in Fig. 15. For cracks of different lengths and the same depth, the longer the crack, the greater the crest trough spacing of the signal. Observing each signal, the abnormal variable of sensor 3 is the largest, so it can be preliminarily concluded that the crack exists near sensor 3. Extract the signal of sensor 3 to obtain the relationship between the signal crest trough spacing and the crack length as shown in Fig. 16. From the results, it can be seen that the length of the crack and the spacing of the signal crest trough have a good linear relationship, which is consistent with the emulation result.

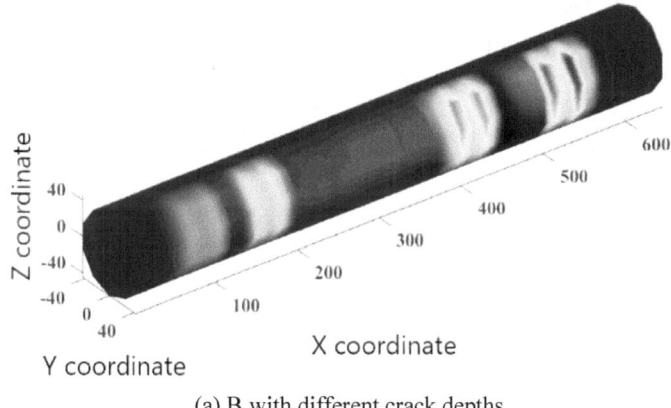

(a) B_z with different crack depths

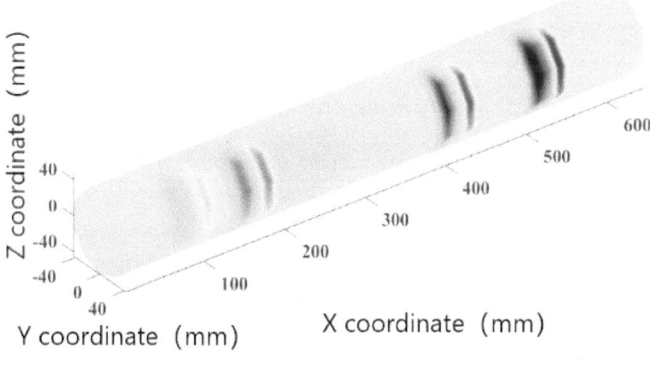

(b) B_r with different crack depths

Fig. 12 Crack detection results at different depths

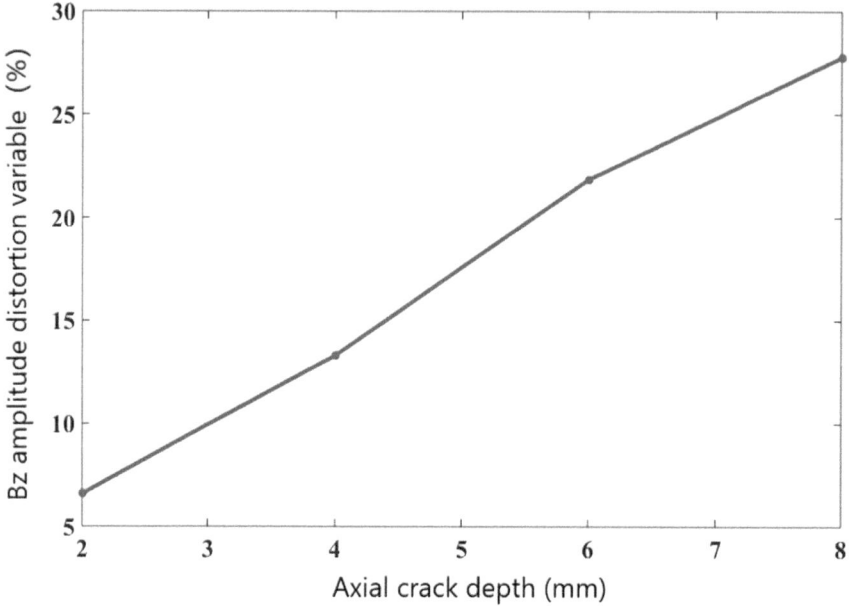

Fig. 13 Bz amplitude distortion of cracks with different depths

crack depth is 0.8mm
the unit is mm

Fig. 14 Pipes of different length crack

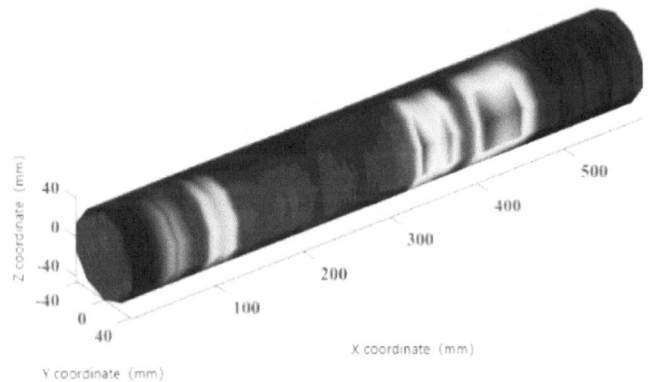

(a) Bz with different crack lengths

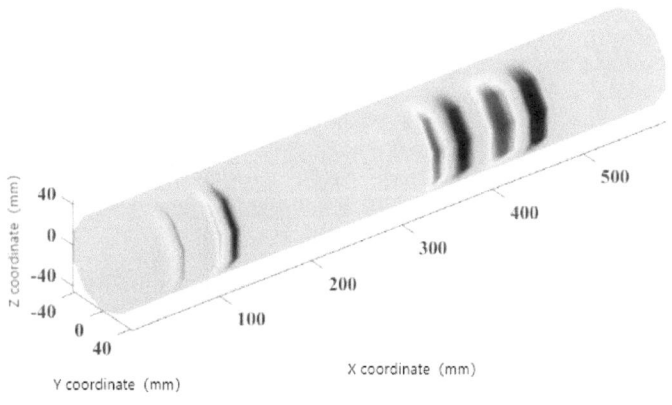

(b) Br with different crack lengths

Fig. 15 Crack detection results at different lengths

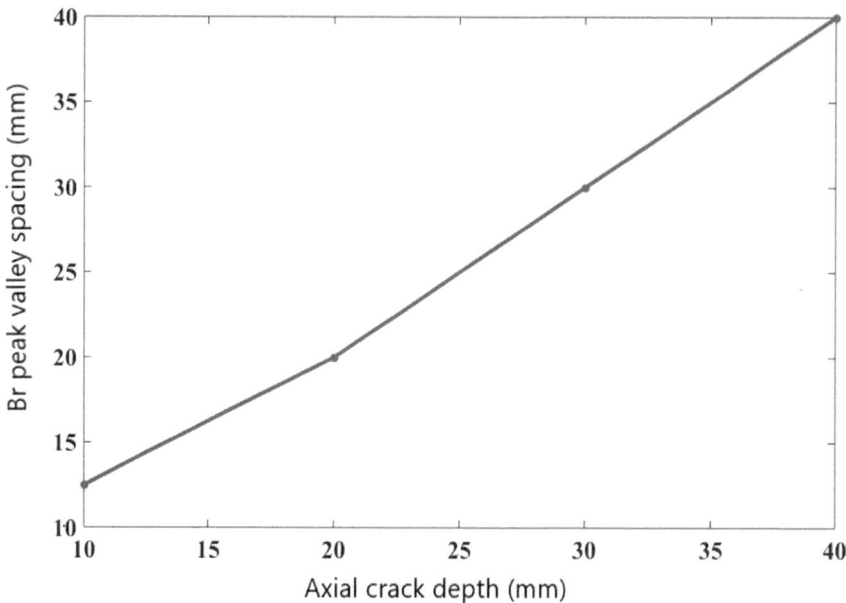

Fig. 16 Br peak-to-valley spacing of cracks of different lengths

3 Conclusion

According to the characteristics of in-pipe detection, based on AC electromagnetic field detection technology, this paper proposes a new type of Hermholtz coil type stainless steel in-pipe detection probe. Through the finite element software Comsol, a uniform electromagnetic field stainless steel in-pipe detection model is established. The distribution and change law of the electromagnetic field in the stainless steel pipeline are analyzed, and the influence of the length and depth of the crack on the characteristic signal is discussed. An experimental system is built to detect the axial cracks of different lengths and depths on the inner wall of the pipeline. The emulation and experimental results show that:

(1) The Hermholtz coil stainless steel pipeline internal detection probe can excite a uniform electric field and magnetic field on the inner surface of the pipeline. When encountering defects, it will cause distortion of the current field, which will further cause distortion of the space magnetic field.

(2) The abnormal variable of the signal contains the depth information of the crack. The deeper the crack depth, the greater the abnormal variable of the signal; the crest trough spacing of the signal contains the length information of the crack. When the crack length is small, the crack length will affect the amplitude and crest trough of the signal. When the crack length is longer, the influence of the crack length on the amplitude of the signal decreases, mainly affecting the spacing of the signal crest trough.

(3) The construction of the Hermholtz coil-type stainless steel pipeline internal detection probe and its experimental system can realize the full-circumferential rapid scanning of the axial cracks on the inner surface of the stainless steel pipeline.

References

1. Zarrinzadeh H, Kabir MZ, Deylami A (2017) Experimental and numerical fatigue crack growth of an aluminium pipe repaired by composite patch. Eng Struct 133:24–32
2. Taherishargh M, Vesenjak M, Belova IV et al (2016) In situ manufacturing and mechanical properties of syntactic foam filled tubes. Mater Des 99:356–368
3. Malekan M, Barros FB, Sheibani E (2016) Thermo-mechanical analysis of a cylindrical tube under internal shock loading using numerical solution. J Braz Soc Mech Sci Eng 38(8):2635–2649
4. Malekan M, Cimini CA Jr (2018) Finite element analysis of a repaired thin-walled aluminum tube containing a longitudinal crack with composite patches under internal dynamic loading. Compos Struct 184:980–1004
5. Canadian Energy Pipeline Association (2007) Stress corrosion cracking recommended practices. Canadian Energy Pipeline Association, Alberta
6. Chen CL (2018) Application of nondestructive testing technology in pressure pipeline. Technol Equip 44(7):80–81
7. Wen Y, Feng Q, Li Y (2014) The application of ultrasonic testing in pipeline welds corrosion. NDT 36(2):50–52
8. Yang LJ, Ge H, Gao SW (2018) Study on high-speed magnetic flux leakage testing technology based on multistage magnetization. Chin J Sci Instrum 39(6):148–156
9. Li W (2007) Research on ACFM based defect intelligent recognition and visualization technique. China University of Petroleum, Dongying
10. Lewis AM, Michael DH, Lugg MC et al (1988) Thin-skin electromagnetic fields around surface-breaking cracks in metals. J Appl Phys 64(8):3777–3784
11. Chacón Muñoz JM, García Márquez FP, Papaelias M (2013) Railroad inspection based on ACFM employing a non-uniform B-spline approach. Mech Syst Signal Process 40(2):605–617
12. Hu XC, Luo FL, He YZ et al (2011) Pulsed alternating current field measurement technique for defect identification and quantification. J Mech Eng 47(4):17–22
13. Li W, Yuan XA, Chen G et al (2016) High sensitivity rotating alternating current field measurement for arbitrary-angle underwater cracks. NDT&E Int 79(2):123–131
14. Li W, Ge J, Wu Y et al (2017) An electromagnetic Helmholtz-coil probe for arbitrary orientation crack detection on the surface of pipeline. Mater Trans 58(4):641–645
15. Ge J, Li W, Chen G et al (2017) Multiple type defect detection in pipe by Helmholtz electromagnetic array probe. NDT&E Int 91:97–107
16. Li W, Yuan XA, Chen GM et al (2017) Research on the detection of surface cracks on drilling riser using the chain alternating current field measurement probe array. J Mech Eng 53(8):8–15
17. Gao YH, Zhang GX, Huang PJ et al (2009) Research on key technology of eddy current testing system based on GMR transducer. Transduc Microsyst Technol 28(11):31–36

Research on the Detection of Surface Cracks on Drilling Riser Using the Chain Alternating Current Field Measurement Probe Array

1 Introduction

In the offshore oil and gas industry, the drilling riser is an essential component that is mostly utilized to link the drilling platform and the seabed wellhead, creating a mud return channel. Stress corrosion cracking (SCC) is a prevalent issue on drilling risers in service because of the complex alternating stress and corrosive conditions. The primary cause of stress corrosion cracking is the collection of microcracks on the riser's surface. These cracks then progressively grow into longer longitudinal cracks when external pressures engage on them, rupturing or leaking the riser and finally resulting in catastrophic drilling mishaps. Therefore, conducting research on the detection of early axial cracks on the surface of the riser is of great significance for preventing riser failure accidents.

Due to the harsh environment of marine drilling and the complex dynamic loads borne by drilling risers, as well as the accumulation of a large amount of debris on the surface, traditional magnetic particle inspection techniques are difficult to detect the surface cracks of the risers. Saturation magnetization is necessary for magnetic flux leakage testing, which has significant requirements for lift-off height, riser surface cleaning, and post-test demagnetization. However, cumbersome processes are not acceptable for high-cost offshore operations.

In this paper, a chain array detection probe based on ACFM is proposed to detect stress corrosion cracks on the outer surface of marine drilling risers. By using multiple U-shaped current-carrying coils, a large uniform field region is excited on the riser pipe surface. The distorted magnetic field above the cracking is extracted by the array GMR sensors, and the cracking characteristic signals are displayed on the computer. The use of a chain structure to encapsulate the excitation coils and sensor arrays enables the detection of metal structures with any diameter, significantly improving detection efficiency and applicability. Moreover, this provides an efficient and universal method for detecting surface cracks on drilling risers.

© The Author(s) 2025

W. Li et al., *Alternating Current Field Measurement Technique for Detection and Measurement of Cracks in Structures*, https://doi.org/10.1007/978-981-97-7255-1_7

2 Principle of Alternative Current Field Measurement

The ACFM technique uses an excitation probe to form a localized uniform current field on the surface of a metallic structure. When the current passes near a defect, it will be deflected, and the deflection current will cause a spatial magnetic field perturbation in which the X-direction flux density Bx appears as a trough in the center of the crack, and the Z-direction flux density Bz appears as opposite peaks at both ends of the crack.

The crack can be quantitatively evaluated by measuring the perturbed magnetic field, with the technical principle as shown in Fig. 1a [1]. Due to the advantages of non-contact measurement (no or little surface cleaning), no calibration, and quantitative assessment, ACFM is very suitable for the detection of surface cracks on underwater structures [2].

The U-type magnetic core current-carrying coil can significantly improve the effect of probe excitation current on metal surfaces [3]. Therefore, in this paper, a rectangular current-carrying coil and a U-shaped Mn–Zn ferrite core are selected to form a U-shaped excitation probe. As shown in Fig. 1b, an alternating electromagnetic field is created within the solenoid when a sinusoidal signal is delivered through the rectangular coil. In the low-frequency excitation state, the Maxwell equations can be used to express the relationship between the electromagnetic fields in space:

$$\nabla \times \boldsymbol{E} = -\frac{\partial \boldsymbol{B}}{\partial t} \tag{1}$$

$$\nabla \times \boldsymbol{H} = \boldsymbol{J} + \frac{\partial \boldsymbol{D}}{\partial t} \tag{2}$$

$$\nabla \cdot \boldsymbol{B} = 0 \tag{3}$$

$$\nabla \cdot \boldsymbol{D} = \rho \tag{4}$$

∇ differential operator.
$\boldsymbol{E}$ electric field strength.
$\boldsymbol{J}$ Current density.
$\boldsymbol{B}$ Magnetic flux density.
$\boldsymbol{H}$ Magnetic field strength.
$\boldsymbol{D}$ Electrical flux density.
ρ volumetric charge density.

According to Maxwell Eqs. (1–4), the Helmholtz equation for the magnetic flux density inside the solenoid can be obtained:

$$\nabla^2 \boldsymbol{B} + k^2 \boldsymbol{B} = 0 \tag{5}$$

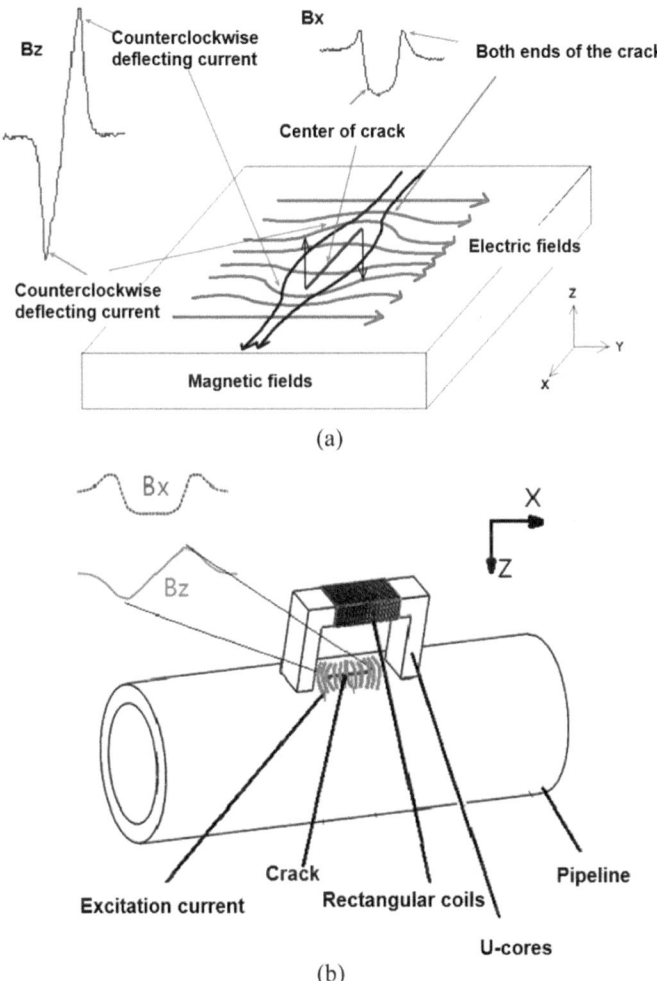

Fig. 1 ACFM schematic diagram. **a** Schematic diagram of ACFM. **b** Schematic diagram of the U-type excitation probe detection

where $k = \omega(u_0\varepsilon_0)^{1/2} = \omega/c$, u_0 is the vacuum permeability, ε_0 is the vacuum capacitance.

In order to facilitate the solution of the governing equation of electromagnetic field at the crack of ACFM, two quantities, vector magnetic potential A and scalar electric potential $\overrightarrow{B} = \nabla \times \overrightarrow{A}$, are introduced to separate the magnetic field variables from the electric field variables, thus making it simple and easy to carry out the numerical solution [4]. The two definitions are as follows:

vector magnetic potential:

$$\overrightarrow{B} = \nabla \times \overrightarrow{A} \tag{6}$$

scalar electric potential:

$$\overrightarrow{E} = -\nabla \phi \tag{7}$$

The vector magnetic potential and the scalar potential defined by (6) and (7) can automatically satisfy the Faraday's law of electromagnetic induction and Gauss's law of magnetic flux. And the partial differential equations of the magnetic field and the electric field, respectively, are obtained after derivation:

$$\nabla^2 \overrightarrow{A} - \mu\varepsilon \partial^2 \overrightarrow{A} \Big/ \partial t^2 = -\mu \overrightarrow{J} \tag{8}$$

$$\nabla^2 \phi - \mu\varepsilon \partial^2 \overrightarrow{\phi} \Big/ \partial t^2 = -\rho \Big/ \varepsilon \tag{9}$$

where $A(X, Y, Z) = A_O(X, Y, Z) + A_P(X, Y, Z)$ is the Laplace operator.

Due to the disturbance of the electromagnetic field caused by the defects, the vector magnetic potential A, which represents the distribution of the magnetic field, can be divided into two parts for ease of calculation:

$$A(X, Y, Z) = A_O(X, Y, Z) + A_P(X, Y, Z) \tag{10}$$

where, A_O is the vector potential function for detecting current induction, and A_P is the vector potential function for electric field disturbance induction caused by defects.

According to electromagnetic induction, the vector potential functions A_O and A_P both satisfy the Laplace equation [5]:

$$\frac{\partial^2 A}{\partial X^2} + \frac{\partial^2 A}{\partial Y^2} + \frac{\partial^2 A}{\partial Z^2} = 0 \tag{11}$$

In the formula, A_O satisfies the boundary condition of the defect-free state:

$$\frac{\partial^2 A_O}{\partial X^2} + \frac{\partial^2 A_O}{\partial Y^2} + \frac{m}{\mu_r} \frac{\partial^2 A_O}{\partial Z^2} = 0 \,|Z = 0 \tag{12}$$

A_P satisfies the boundary condition of the defect-free state:

$$\frac{\partial^2 A_p}{\partial X^2} + \frac{\partial^2 A_p}{\partial Y^2} + \frac{m}{\mu_r} \frac{\partial^2 A_p}{\partial Z^2}$$
$$= \left(2 + \frac{cm}{u_r}\right) \frac{\partial A_p}{\partial Z} \delta(Y) | Z = 0 \tag{13}$$

where $m^2 = 2i/\delta^2$, δ is the skin depth, c is the defect width [6]:

$$\delta = (\mu_r\mu_0\pi\sigma f)^{-1/2} = (2/\sigma\mu\omega)^{1/2} \tag{14}$$

where μ_r is the relative permeability of the material, μ is the permeability of the material, $\mu = \mu_r\mu_0$, σ represents the conductivity, f represents the current frequency, and ω represents the angular frequency of the current.

When the uniform current field excited by the U-type current-carrying coil passes through the defective region, the current gathers at the two ends of the crack, which causes the pipe radial spatial magnetic field Bz to produce peak and valley due to the opposite direction of the current deflection, and the distance between the peak and valley reflects the length of the crack. At the same time, due to the effect of crack depth, the current density gradually decreases in the depth direction of the crack center, which will produce a decrease in the magnetic flux density Bx in the axial direction of the pipe, forming a wave valley. The larger the crack depth is, the larger the Bx trough depth (the distortion rate of the amplitude in the background magnetic field) is, so the Bx signal contains crack depth information. Owing to the accurate ACFM mathematical model, the crack length and depth information can be obtained by applying the uniform electromagnetic field to detect and evaluate the defects.

3   Finite Element Method Model

3.1   Modeling Design

The riser [7] that is frequently used in the Nanhai 8 platform, with an outside diameter of 533.4 mm (21 inches) and a wall thickness of 15.875 mm (5/8 inch), is chosen as the research object for this paper. Table 1 displays the precise dimensions and specifications.

The ACFM simulation model for the axial cracks on the riser surface is established by using ANSYS finite element software, as shown in Table 2. In order to reduce the number of computational units, the model is built only in the upper part with the cracked region. The dimensions of the simulation model are shown in Table 2. The excitation coil is a U-shaped core current-carrying coil [8] and the excitation signal is a sinusoidal signal with a frequency of 6 kHz and an amplitude of 1 V [9, 10]. The current density around the crack is extracted in the simulation model as shown in Fig. 2b. It can be clearly seen that the U-shaped magnetic core excites two eddy current zones on the surface of the riser, and the middle of the vortex

Table 1 Dimensions of riser

Model	Diameter/mm	Wall thickness/mm	Length/mm	Material
Dimensions	533.4	15.875	18.21	X80 steel

Table 2 Dimensions of simulation model

Model	Diameter/mm	Wall thickness/mm	Length/mm	Length of crack/mm	Width of crack/mm	Depth of crack/mm
Dimensions	533.4	15.875	2000	45	0.8	5

region approximately forms a uniform electric field region. There is a high current density at both ends of the crack because the uniform electric field gathers at both ends as it travels through the crack. At the same time, there will be a sparse current density inside the middle crack air gap as a result of the current being diverted to the depth direction when it passes through the crack's center. It is stated that perturbing the electric field will result in a change in the spatial magnetic field based on the Maxwell classical electromagnetism relations. A path is defined along the top of the crack with the path step set to 100, and the magnetic flux densities Bx (axial direction) and Bz (radial direction) above the crack are extracted. As seen in Fig. 2c, Bz produces a peak and valley at the ends of the crack, respectively, and its peak and valley starting position reflects the crack length. Bx produces a wave peak at the center of the crack (the negative sign indicates the direction, and its absolute value is the valley). Bx contains the information of crack depth due to the relationship between the flux density attenuation and the crack depth. The aforementioned research demonstrates that the simulation model developed in this work may achieve the quantitative detection of cracks on the riser surface while adhering to the ACFM detection law.

3.2 Excitation Probe Spacing Simulation Analysis

Determining the distance between the row of excitation coils is necessary for designing a chain array probe. Excitation coil spacing should be such that the column surface in the surrounding area is fully covered, but it shouldn't be too small to prevent superposition of the excitation zone and space wastage. With the use of the simulation model, as indicated in Fig. 3a, the core is offset from the crack by a specific arc length L. The Bx and Bz flux densities above the crack are then extracted to provide the Bx and Bz flux density distortions at various offset distances, as indicated in Fig. 3b. The extracted Bx and Bz flux density distortions are shown in Table 3.

Figure 3 shows that both the Bx and Bz amplitudes decrease significantly as the offset distance of the excitation core increases. The Bz magnitude distortion decreases linearly when the core is shifted by 20 mm; it tends to decrease slowly when the core is shifted by more than 20 mm; and when the core is offset by 50 mm, the Bz magnitude distortion is 11.23% of the Bz magnitude aberration when the core is located directly above the crack (offset angle of 0°). The Bx amplitude distortion falls sharply and then tends to decrease steadily when the core is offset by 10 mm. When the offset distance reaches 50 mm, the Bx amplitude distortion

Fig. 2 Simulation result.
a Simulation model of the
riser surface axial crack.
b Distortion of surface
current in the crack region.
c Extractions of the crack
characteristic signal

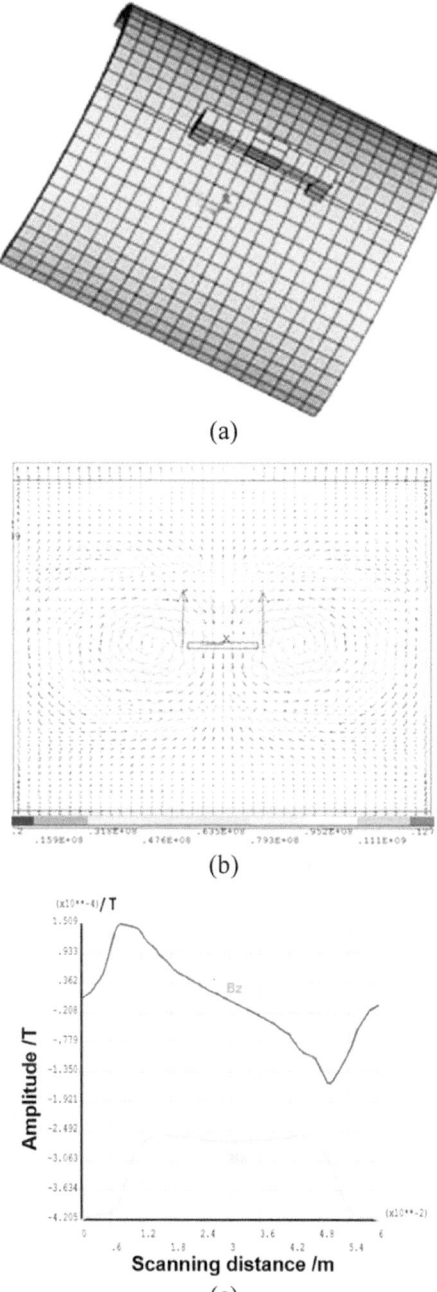

(a)

(b)

(c)

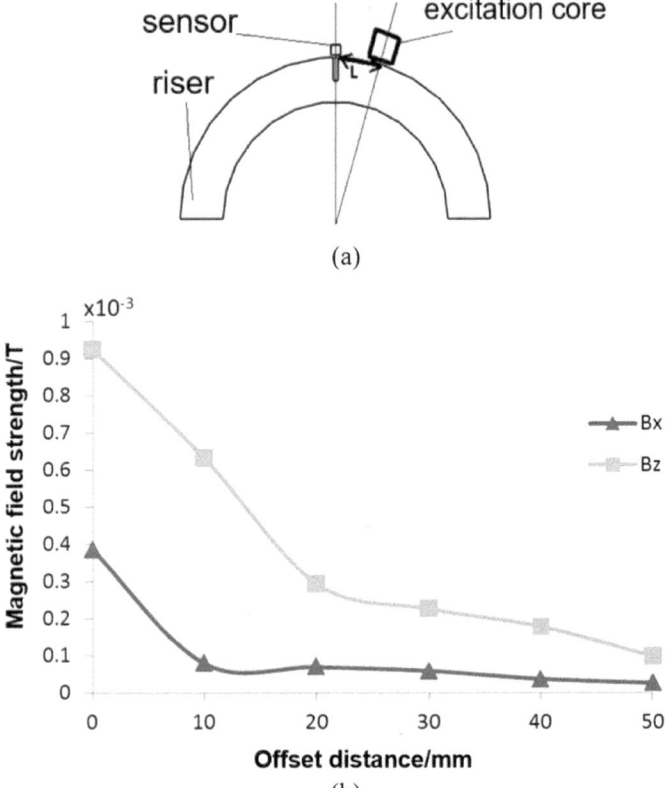

(a)

(b)

Fig. 3 Analysis of the influence region of an excitation probe. **a** Excitation core offset distance diagram. **b** Bx and Bz flux density distortion plots

Table 3 Bx and Bz flux density distortions

L/mm	0	10	20	30	40	50
$\Delta Bz/ \times 10^{-4}$T	9.44	6.28	2.95	2.11	1.77	1.06
$\Delta Bx/ \times 10^{-4}$T	3.85	0.74	0.72	0.65	0.42	0.24

is 6.23% of that when the core is located directly above the crack (offset 0°), and the distortion amplitude is less than 10% of the normal detection position. Such distortion amplitude makes it easy to form an interfering signal, which results in leakage detection and false detection. Therefore, according to the simulation results, the maximum distance of the excitation core offset from the crack can be set to 40 mm, i.e., the maximum design spacing between the excitation cores of the chain array probe is 80 mm.

3.3 Sensor Spacing Simulation Analysis

After determining the chain excitation core spacing, it is also necessary to determine the design spacing of the sensors below the core. With the help of the simulation model, the spatial flux densities Bx and Bz on the surface of the riser are extracted at a lift-off height of 4 mm above the crack, as shown in Fig. 4a, b. It can be seen that the flux density distortion region above the crack is distributed between − 5 and 5 mm in the crack width direction.

As shown in Fig. 5a, the excitation core is located directly above the crack, and the Bx and Bz flux densities are extracted at the positions of the sensor with different offset crack distances (0–6 mm). Figure 5b plots the Bx and Bz flux density distortions at various offset positions, and Table 4 displays the flux density distortions. It is evident that the Bx and Bz amplitude distortions decrease with the increase of the sensor offset crack distance. When the offset distance reaches 6 mm, the amplitude distortion of Bz reaches the order of 10^{-5} T, and the characteristic signal is already very weak. In order to obtain the characteristic signals of the defects, the maximum offset crack distance of the detection sensor is 5 mm, or the maximum design spacing of the sensor below the core is 10 mm.

4 Experimental Research

4.1 Design of the Testing System

Based on the principle of ACFM, a drilling riser chain array detection system is designed as shown in Fig. 6. The current-carrying coil of the U-shaped magnetic core receives a sinusoidal excitation signal from the signal generator with an amplitude of 1 V and a frequency of 6 kHz. The riser's surface is excited by a uniform current field from the excitation core, and when this excitation electric field gets through the surface cracks, the perturbation current distorts the spatial magnetic field. After signal conditioning (amplification and filtering), the spatial magnetic field distortion signals are detected by the array sensors placed at the bottom of the chain probe and sent to the acquisition card. The array probe data is transferred by the acquisition card to a computer for defect analysis and signal display.

4.2 Sensor Circuit Design

The sensitivity, large linear range, small size, and other characteristics of the Giant Magneto Resistance (GMR) magnetic field sensor make it an excellent choice for weak magnetic field detection [11]. GMR magnetic field sensors are used in the drilling riser chain array ACFM probe sensors designed in this paper. In order to

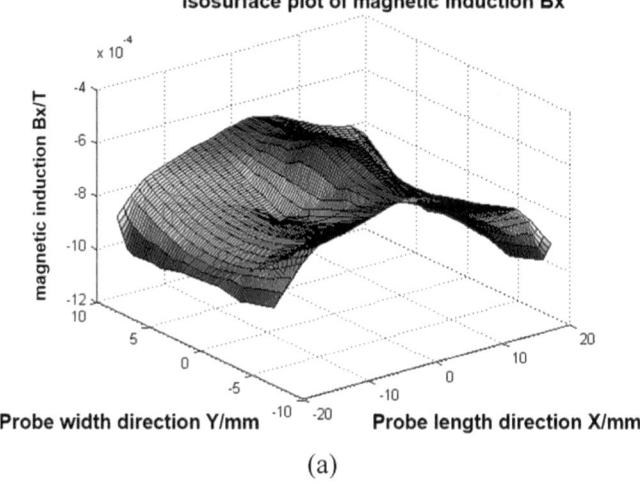

(a)

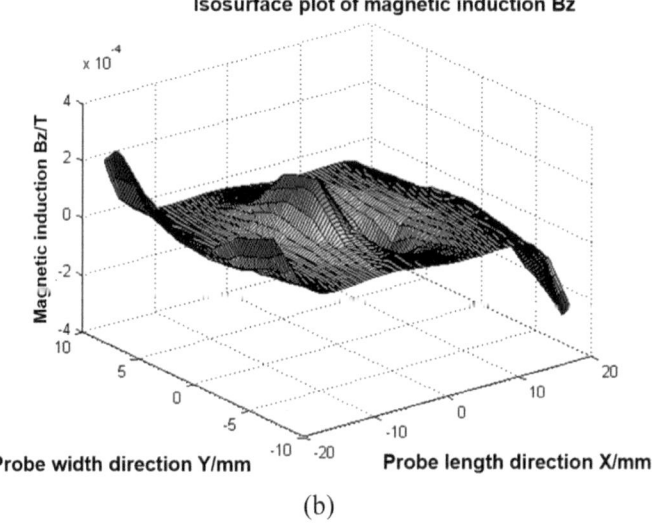

(b)

Fig. 4 The spatial flux densities Bx and Bz around the crack. **a** *Bx* space flux density. **b** *Bz* space flux density

measure the Bx and Bz signals, the GMR sensors are orthogonally welded on both sides of the PCB circuit board. Besides, the common ground capacitance is utilized for filtering, noise reduction, and anti-interference, while the AD620 amplifies the signals [12]. The front and back sides of the sensors' circuit are shown in Fig. 7a, b.

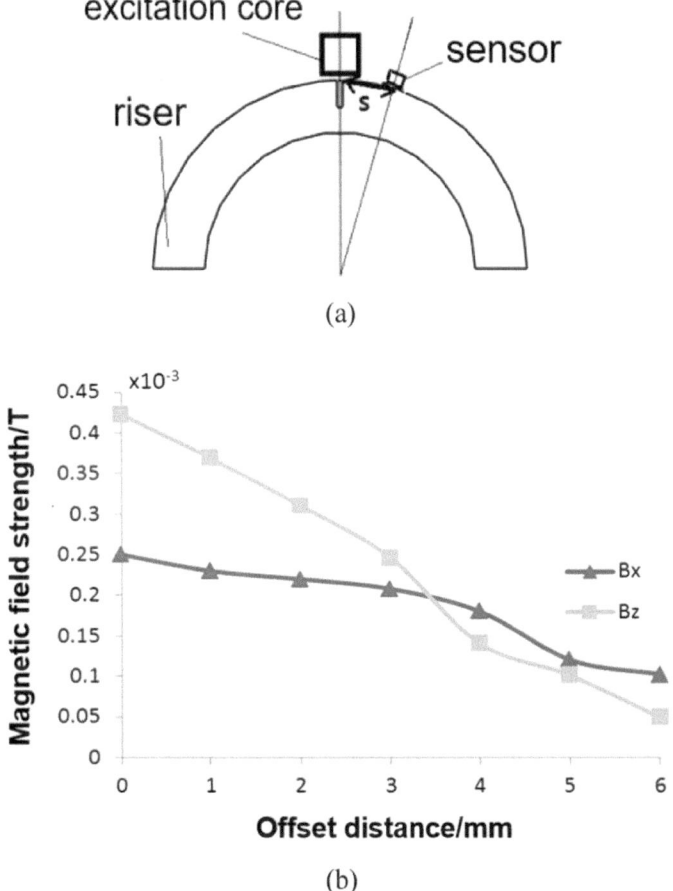

(a)

(b)

Fig. 5 Area of flux density distortion in the cracked area. **a** Diagram of the detecting sensor offset distance. **b** Bx and Bz flux density distortions

Table 4 Magnetic flux density distortions around crack

L/mm	0	1	2	3	4	5	6
$\Delta Bz / \times 10^{-4}T$	4.23	3.69	3.10	2.46	1.41	1.01	0.05
$\Delta Bx / \times 10^{-4}T$	2.51	2.30	2.22	2.08	1.81	1.20	1.02

4.3 Chain Probe Design

The excitation core is fixed in the center of a single link, and the GMR detection circuitry is distributed on the bottom plate of the single link, keeping the lift-off height of the GMR sensors at 4 mm. The individual links can be freely bent to

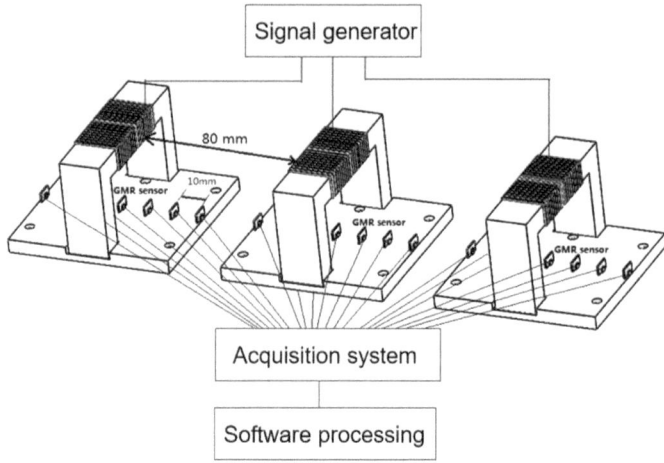

Fig. 6 Diagram of the testing system design

each other to ensure that the arc crack detection of variable pipe diameter can be realized, and the link can be freely disassembled according to the detection area and space restrictions. Due to the limitations of the acquisition card, the experimental system adopts three single links, each of which has a current-carrying excitation core inserted. 21 uniformly numbered detecting circuit boards are dispersed beneath the links (No. 0–20 sensor). Figure 8a depicts the probe design, and Fig. 8b shows the chain probe after it has been assembled and processed.

4.4 Test Experiment

The test specimen is a riser with an outer diameter of 533.4 mm and a wall thickness of 15.875 mm. Its outer surface has an EDM-machined axial crack (Crack No. 0) with dimensions of 45 mm in length, 0.8 mm in width, and 5 mm in depth. Furthermore, there is consistency between the computer model and the experimental parameters. Figure 9a displays the specimen and the crack, and Fig. 9b shows the entire inspection devices. The chain probe is mounted on the spindle of the bench, and the computer controls the bench to telescope or rotate via PLC. A signal generator provides a sinusoidal signal with a frequency of 6 kHz and an amplitude of 1 V to the chain probe. The defect signals measured by the sensors inside the chain probe are processed and captured by the signal processor and delivered to the computer. Intelligent visualization software based on LabVIEW and MATLAB inside the computer displays the characteristic signals of the defects.

The chain probe is placed close to the surface of the riser. Using the computer to control the bench to extend slowly at a uniform speed of 5 mm/s, the bench drives the chain probe to sweep uniformly across the arc along the axial direction of the

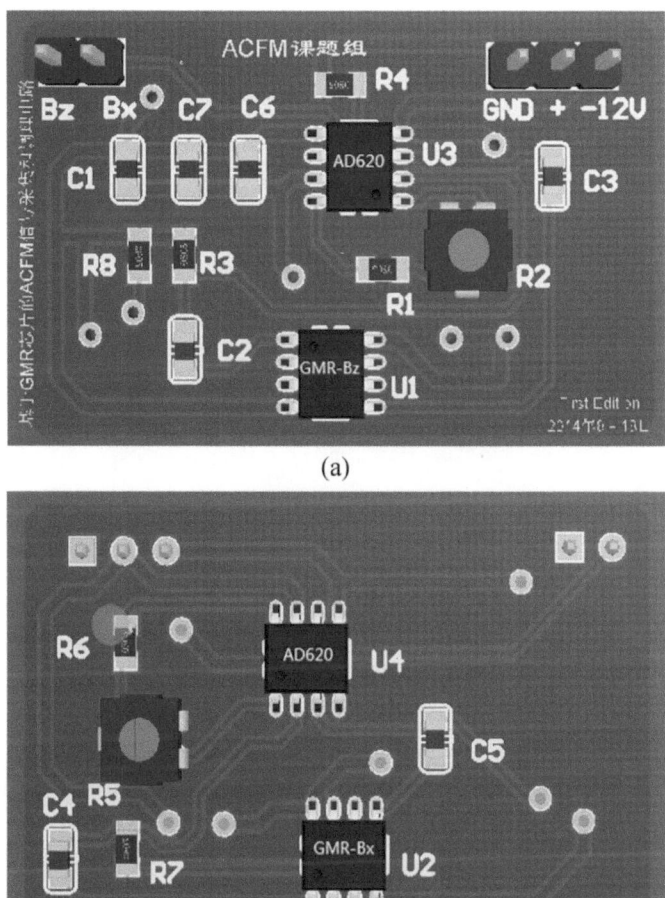

(a)

(b)

Fig. 7 Circuit design for detection sensors. **a** Front side of PCB (Bz sensor). **b** Backside of PCB (Bx sensor)

riser, and the GMR sensors pick up the Bx and Bz magnetic flux densities above the crack and implement the A/D conversion by the acquisition card. The intelligent visualization software inside the computer stores and recalls the Bx and Bz signals of each channel for display. Finally, the magnetic flux densities above the crack of the inspection board test are shown in Fig. 10.

Similarly, another axial crack (crack No. 1) on the surface of the laboratory riser machined by EDM of 40 mm long, 0.8 mm wide and 5 mm deep is detected using a chain probe at a rate of 5 mm per second. The test results with the most obvious amplitude of the characteristic signal are selected, as shown in Fig. 11.

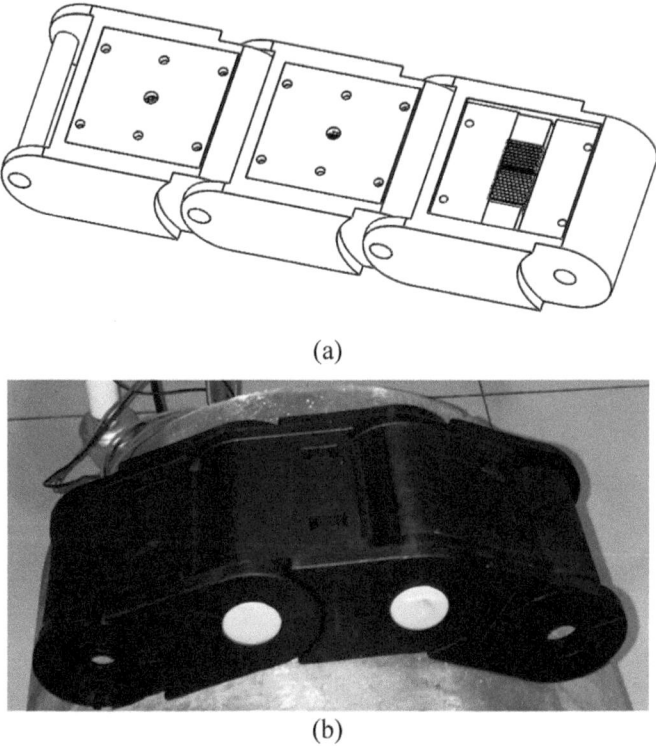

(a)

(b)

Fig. 8 Chain probe design. **a** Design drawing of the chain probe. **b** Physical drawing of the probe design

4.5 Analysis of Experimental Results

The Bx and Bz measured by sensors No. 9–11 show some distortions, with a trough in Bx and two peaks in Bz. Since the GMR sensor measurement value is the absolute value of the magnetic field at the location, the internal program of the software adopts the root-mean-square algorithm for the collected data, so the collected Bz signal changes from the original peak and valley to two peaks. The distance between the two peaks of Bz reflects the length of the crack, and the depth of the valley of Bx reflects the depth of the crack. The aberration pattern of the Bx and Bz signals is in agreement with the results of the simulation model, which is in line with the principle of ACFM detection.

At the same time, it can be seen that the Bx and Bz distortions measured by sensor No. 10 have the strongest signals, while the Bz signal measured by sensor No. 11 and No. 9 adjacent to sensor No. 10 has weak characteristics and distortion phenomena. Moreover, the signals measured by the No. 8 and No. 12 sensors are cluttered, and no characteristic signal of cracks can be seen. From this, it can be judged that the crack appears near the No. 10 sensor, so as to realize the crack location detection.

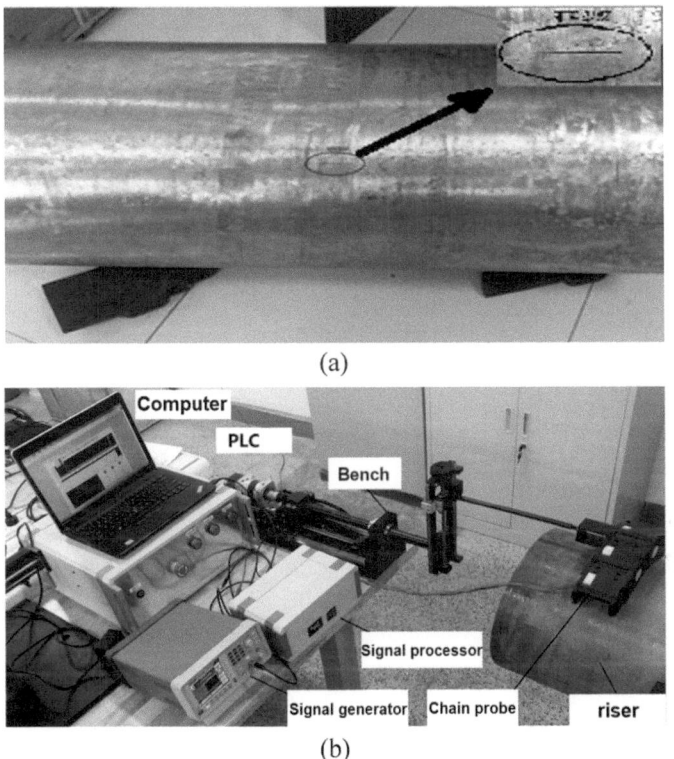

(a)

(b)

Fig. 9 Specimen and testing system. **a** Riser specimen and axial crack. **b** The testing system

From Fig. 10c, it can be obtained that the peak sampling point interval of the characteristic signal Bz of crack No. 0 is 1255–1755, and the number of sampling points is 500. Figure 11 shows the results of another crack No. 1 of the same depth and width tested by the chain probe at the same test speed. The characteristic signal Bz interval for this crack is 1095–1520 with 425 sampling points. Since the same speed is used, the predicted crack length can be obtained from the ratio of the number of sampling points of the Bz characteristic signal of crack No. 1 to the number of sampling points of crack No. 0 as shown in Eq. (15).

$$L_1 = L_0 \times N_1/N_0 \tag{15}$$

L_0 is the length of crack No. 0, N_0 is the number of peak sampling points of the feature signal of crack No. 0, L_1 is the predicted crack length, and N_1 is the number of peak sampling points of the crack sign signal of crack No. 1.

As predicted by formula (15), the length of crack No. 1 is 38.25 mm, and the quantitative error of crack length is 4.36%, which means the detection accuracy meets the engineering requirements.

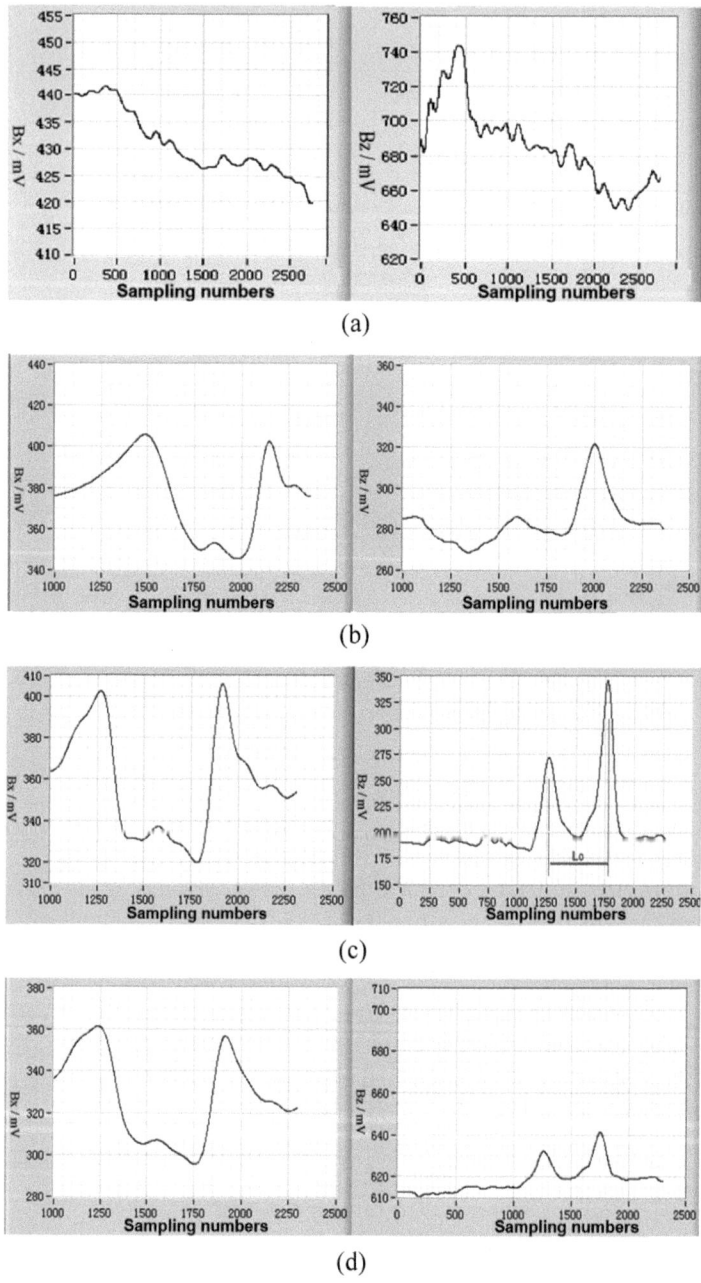

Fig. 10 Test result of 45 mm crack detection of drilling riser. **a** Sensor no. 8 test results. **b** *Sensor no. 9 test results.* **c** Sensor no. 10 test results. **d** Sensor no. 11 test results. **e** Sensor no. 12 test results

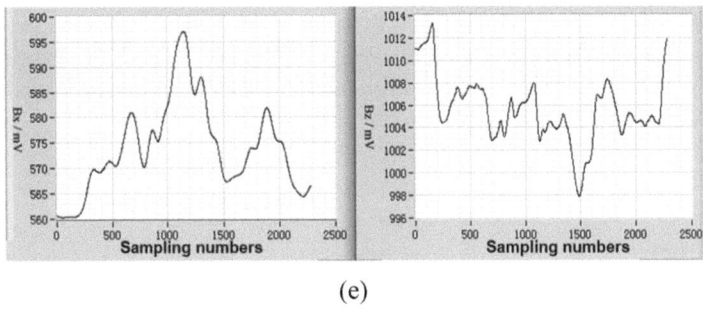

(e)

Fig. 10 (continued)

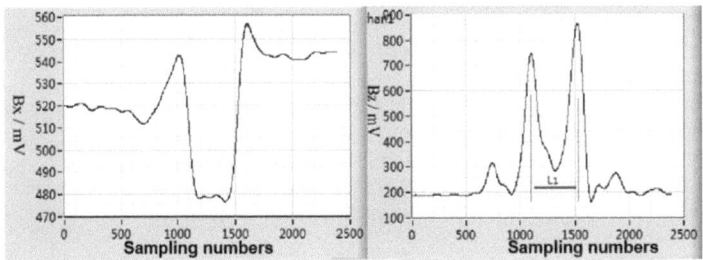

Fig. 11 Test result of 45 mm crack detection of riser

5 Conclusions

In this paper, a chain array probe based on the ACFM technology principle is created for the characteristics of stress corrosion cracks on the drilling riser surface. This probe can accomplish rapid, quantitative, and location-based crack surface measurement. The ACFM simulation model of a U-shaped current-carrying coil is established with the aid of ANSYS software in order to analyze the characteristic signals of the surface cracks of the riser, and optimize the distances of the excitation cores and sensors. Based on the simulation results and GMR sensors, a chain-array ACFM probe is constructed, and further tests are carried out to find the riser's surface cracks. Results from experiments and simulations indicate that:

(1) U-type ACFM current-carrying coil excitation probe can excite a uniform electric field region on the surface of drilling riser.Axial stress corrosion crack on the riser's surface can disturb the electric field, which in turn causes the spatial magnetic field to become aberrated due to the perturbation current. Accordingly, the length and depth of the crack is contained in the aberration magnetic fields Bx and Bz.

(2) In order to realize the thorough detection of cracks on the surface of the drilling riser, the design spacing of the excitation cores should be no more than 80 mm, and the design spacing of the detection sensors should be no more than 10 mm.

(3) The chain array probe designed by using GMR sensors can effectively detect cracks on the surface of the riser, realize rapid, quantitative and location detection, and provide a set of effective methods for early crack discovery and integrity management of drilling riser.

References

1. Li W, Chen G, Zhang C et al (2013) Simulation analysis and experimental study of defect detection underwater by ACFM probe. Chin Ocean Eng Soc 27(2):277–282
2. Amineh RK, Ravan M, Sadeghi SHH et al (2007) Using AC field measurement data at an arbitrary lift off distance to size long surface-breaking cracks in ferrous metals. NDT&E Int 40(7):537–544
3. Li W, Chen G (2013) Simulation analysis of U-shape inducer for ACFM. J Syst Simul 40(6):738–742
4. Li W (2007) Research on ACFM based defect intelligent recognition and visualization technique. Dongying China Univ Petrol
5. Kang Z, Luo F, Hu Y et al (2004) The mathematical modelling of alternating current field measurement and the establishment of approximately uniform field. NDT 26(11):546–550
6. Tao M, Chen D, Lu Q et al (2012) Eddy current losses of giant magnetostrictors: modeling and experimental analysis. J Mech Eng 48(13):146–151
7. Liu X, Chen G, Chang Y et al (2013) Analyses and countermeasures of deepwater drilling riser grounding accidents under typhoon conditions. Petrol Explor Dev 40(6):738–742
8. Li W, Chen G (2009) Defect visualization for alternating current field measurement based on the double U-shape inducer array. J Mech Eng 45(9):233–237
9. Li W, Chen G, Li W et al (2011) Analysis of the inducing frequency of a U-shaped ACFM system. NDT&E Int 44:324–328
10. Li W, Chen G, Yin X et al (2013) Analysis of the lift off effect of a U shaped ACFM system NDT&E Int 53:31–35
11. Gao Y, Zhang G, Huang P et al (2009) Research on key technology of eddy current testing system based on GMR transducer. Transduc Microsyst Technol 28(11):31–36
12. Liu Y, Yang S, Yu B (2011) Detection of crack om nonferromugnetic metal using eddy current test based on GMR. J Vibr Measur Diagn 31(6):747–753

An Electromagnetic Helmholtz-Coil Probe for Arbitrary Orientation Crack Detection on the Surface of Pipeline

1 Introduction

Pipelines provide the safest and most economical form of transportation of crude oil, natural gas, and other petrochemical commodities compared to truck, rail cars, and tankers [1]. Suffering from the degeneration of materials, for example crack, corrosion, it is important to carry out nondestructive testing (NDT) to maintain high reliability of pipeline transportation.

Surface cracks, especially the stress corrosion crack (SCC), are one of the most harmful degradations considering their effect on structural integrity [2]. Subjected to hoop stress and land movement load, the orientation of surface cracks may be axial direction, circumferential direction or the others. Ultrasonic testing is a widely used technique in pipeline detection. However, the need of couplant limits the detection in gas transportation pipeline [3, 4]. As to magnetic flux leakage, the leaking magnetic flux is proportional to the opening of the crack, so it is not sensitive to axial tight SCC [5−7]. Magnetic particle inspection (MPI) is the most reliable NDT method in pipeline detection. While MPI is a very effective inspection technique, its field use can be costly when one considers the surface cleaning and operating time [5]. The eddy current (EC) could not make accurate size assessment of cracks [8].

The aim of this paper is to propose an electromagnetic Helmholtz-coil probe for oriented cracks mapping and sizing on the surface of pipeline. This paper is organized as follows. In Sect. 2, the model of the Helmholtz-coil probe is presented and the perturbation of the magnetic field above the crack is analyzed through the finite element software COMSOL. In Sect. 3 the electromagnetic Helmholtz-coil probe is set up and the oriented crack in the range of 0°–90° are mapped through experimental testing.

© The Author(s) 2025

W. Li et al., *Alternating Current Field Measurement Technique for Detection and Measurement of Cracks in Structures*, https://doi.org/10.1007/978-981-97-7255-1_8

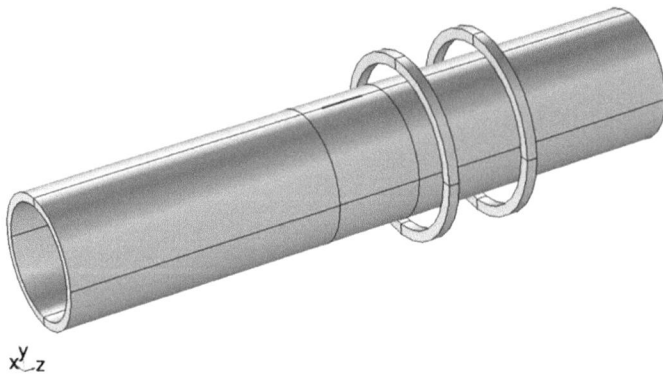

Fig. 1 FEM model the electromagnetic Helmholtz-coil probe

2   Simulation of the Electromagnetic Helmholtz-Coil Probe

An electromagnetic Helmholtz-coil probe consists of two critical components was presented in this paper: a Helmholtz-coil excitation and a detecting sensor array. The Helmholtz-coil excited by alternating current (AC) was coaxial with the pipeline to induce a uniform current field along the circumference of the pipeline and a uniform magnetic field along the axis of the pipeline [9]. The tunnel magneto resistive (TMR) detecting sensor array was used to measure the magnetic field signals.

2.1   Finite Element Model of the Electromagnetic Helmholtz-Coil Probe

The finite element model of the electromagnetic Helmholtz-coil probe was built through the COMSOL software, as shown in Fig. 1. The "magnetic field physics" was selected and the "impedance boundary condition" was applied on the surface of the pipeline. The dimensions of the model are shown in Table 1 and the characteristic parameters are shown in Table 2. The electric current and magnetic flux density distribution on the surface of pipeline without crack are shown in Fig. 2. As shown in Fig. 2, the uniform electric current field and magnetic field are induced on the surface of pipeline [10, 11].

2.2   Result Analysis

In the simulation, the oriented cracks in the range of 0°–90°, as shown in Fig. 3, were detected. The position of the probe was moved along the axial direction (z-axis) to

Table 1 The size of the FEM model

Model	Diameter/mm	Length/mm	Width/mm	Depth/mm	Interval/mm
Pipeline (D/d)	65/45	300			
Helmholtz-coil	80	4			40
Crack		20	1	5	

Table 2 The characteristic parameter of the FEM model

Number of turns	Pipeline material	Conductivity/S/ m	Permeability	Excitation current/A	Frequency/Hz
500	Carbon steel	1.12e7	4000	1	1000

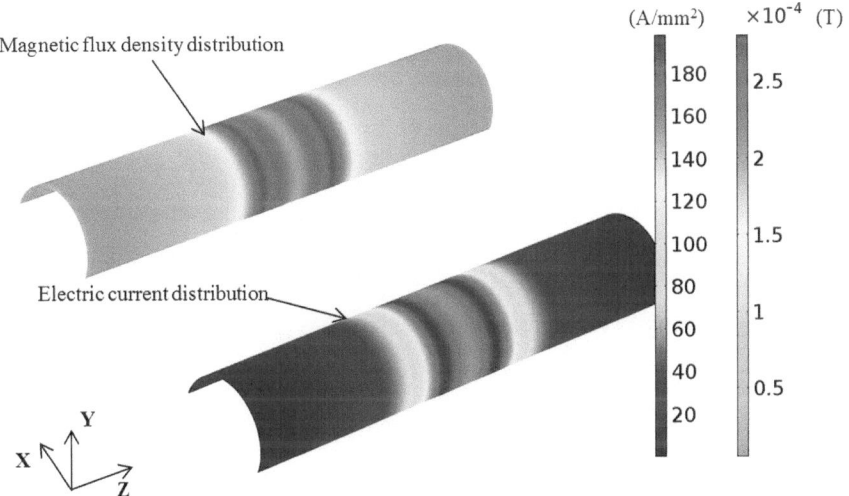

Fig. 2 Electric current and magnetic flux density distribution on the surface of the pipeline

simulate the probe scanning process along a pipeline during experimental studies. The response magnetic field of the probe was measured at a 1 mm lift-off. Two components of the magnetic field, axial direction (z-axis) Bz and radial direction (y-axis) By, were measured as the characteristic signals, as shown in Fig. 4.

As shown in Fig. 4a, b, with the increasing of the angle, the Bz signal changes from dip to peak gradually and the By signal remained a peak and dip. However, the order of the peak and dip in By signals varies with the increasing of crack orientation.

The electric current density around the 0° orientation crack on the surface of pipeline and magnetic flux density around the 90° orientation crack in the x-y plane of the model are shown in Fig. 5.

Fig. 3 Oriented cracks in the range of 0°–90°

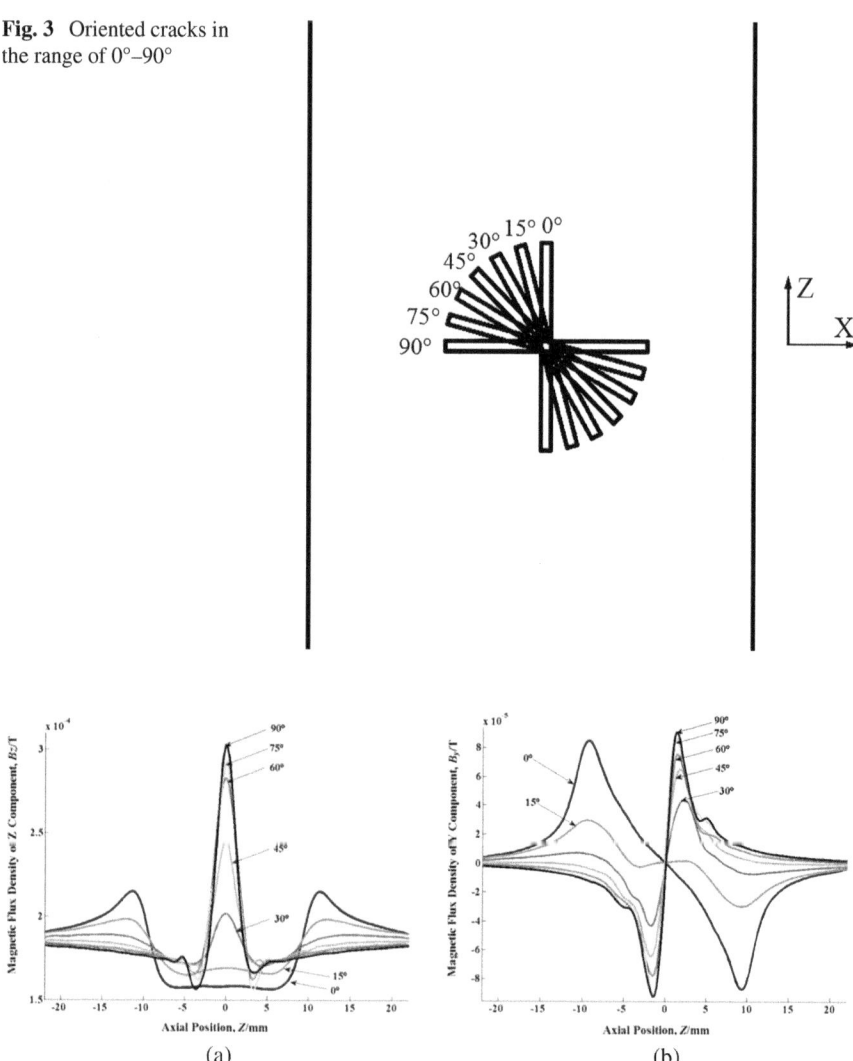

Fig. 4 Bz and By signals, **a** Bz, **b** By

It can be seen from the Fig. 5a that the electric current concentrating in the tips of the crack, which is expressed as arrow, when the crack of 0° orientation is detected is the electric current perturbation effect caused by discontinuous conductivity. As shown in Fig. 5b, the magnetic flux leaks in the edge of the crack, which is expressed as line, when the crack of 90° orientation is detected is the magnetic flux leakage effect caused by discontinuous permeability. These phenomenon also explain the variations of B_z and B_y signal in Fig. 4 that the B_z and B_y signals of the crack of 0°, 15° orientation are caused by the electric current perturbation effect and the signals

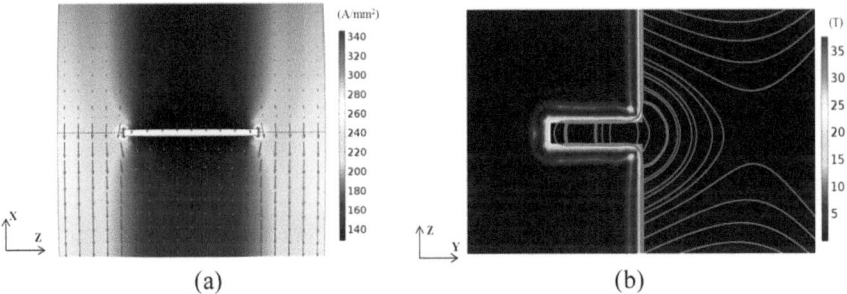

Fig. 5 The electric current distribution and magnetic flux density around the crack, **a** electric current density around the 0° orientation crack on the surface of pipeline, **b** magnetic flux density around the 90° orientation crack in the x-y plane of the model

of the crack in the range of 30°–90° orientation are caused by the magnetic flux leakage effect.

An implication of this is the possibility that the combination effect of the electric current perturbation and magnetic flux leakage around the crack may have the advantage of detecting the oriented crack on the surface of pipeline.

3 Experiments

3.1 System Set Up

The structure of the Helmholtz probe is shown in Fig. 6. The excitation coils were wound on the polymer frame. Detecting sensors were equally-spaced installed in the polymer frame. Supports were used to fix the probe and keep the constant lift-off to different pipeline. The extension-type tape was used to fix the sensor array. A sensor array containing Tunnel Magneto Resistive (TMR) was employed to scan the full circumference of the pipeline. Considering the space of the Helmholtz-coil electromagnetic probe and actual manufacturing difficulty, a 24 equal-spaced sensor array was selected. An electromagnetic Helmholtz-coil detection system was built, as shown in Fig. 7. The excitation source produced an alternating current signal with the frequency of 1 kHz and the magnitude of 10 V. The turns of the coil were 1000 in total. The current was transferred to the excitation coil through the power amplifiers. The detecting sensor array picked up the magnetic field and translated it into electric signal. The signals were amplified and filtered in signal processing module. And then, the signals were converted into digital signal by an A/D convertor and sent to PC for signal processing. A detection software was developed to achieve defects recognition. The Bz and By signals were shown in the display screen of computer. The Helmholtz-coil probe was fixed in an axial scan table, as shown in Fig. 7.

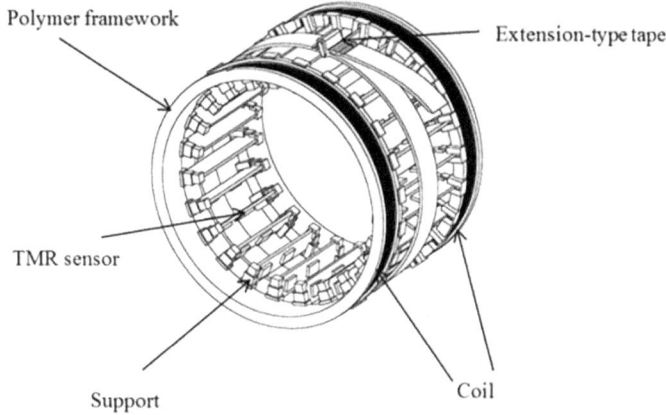

Fig. 6 Structure of Helmholtz-coil probe

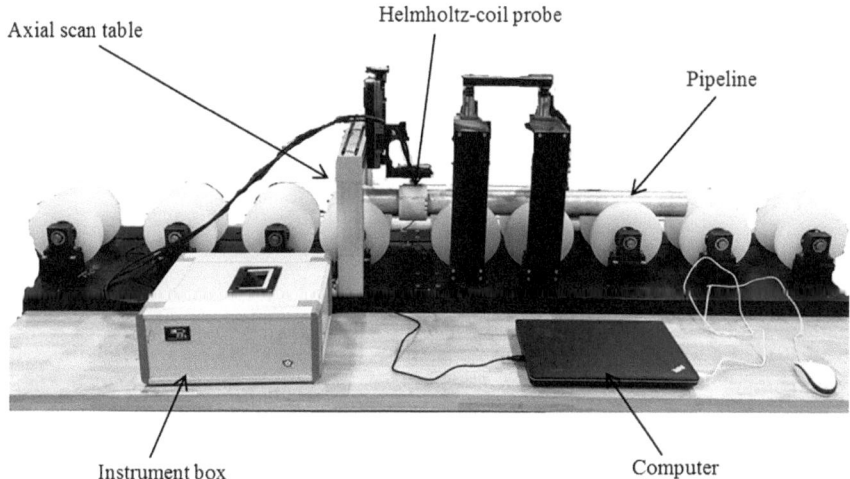

Fig. 7 Experimental setup

3.2 Result Analysis

The pipeline made from carbon steel was detected to test the detectability of Helmholtz-coil probe to crack. The axial cracks, which are 1 mm width and 30 mm length, with the depth in the range of 2–10 mm were machined on the surface of pipeline by Electric Discharge Machining (EDM), as shown in Fig. 8.

The pipe string was moved along the axial direction at a speed of 10 mm/s and the Bz and By signals were measured and shown in Fig. 9. It can be seen from the figures that all of the cracks on the surface of pipeline could be detected obviously.

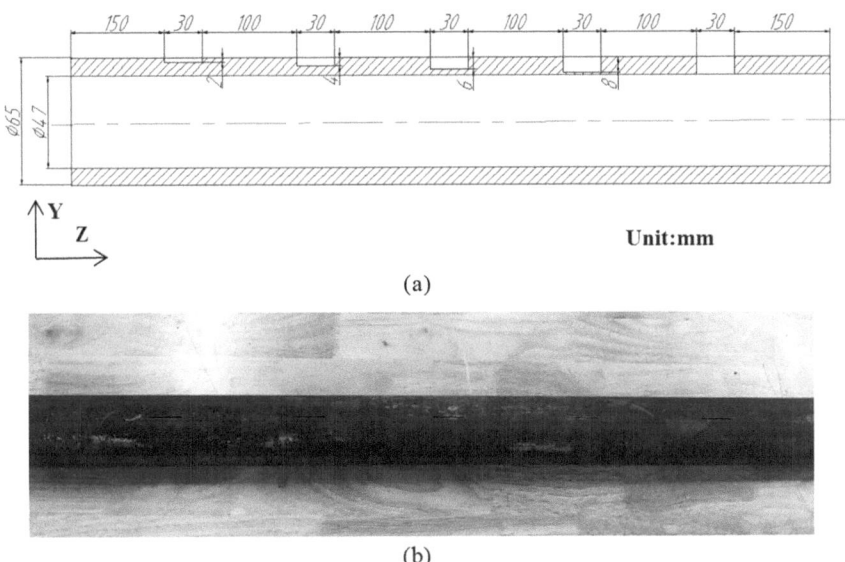

(a)

(b)

Fig. 8 **a** Schematic diagram of pipeline and **b** actual cracks in pipe strings

Furthermore, the length of cracks can be obtained accurately. The amplitudes of the signal are proportional to the depth of the cracks.

Oriented cracks with the orientation in the range of 0°–90° were machined on the surface of pipeline by EDM, as shown in Fig. 10, and the parameters of the cracks were 45 mm × 1 mm × 5 mm (length × width × depth). The pipeline was scanned

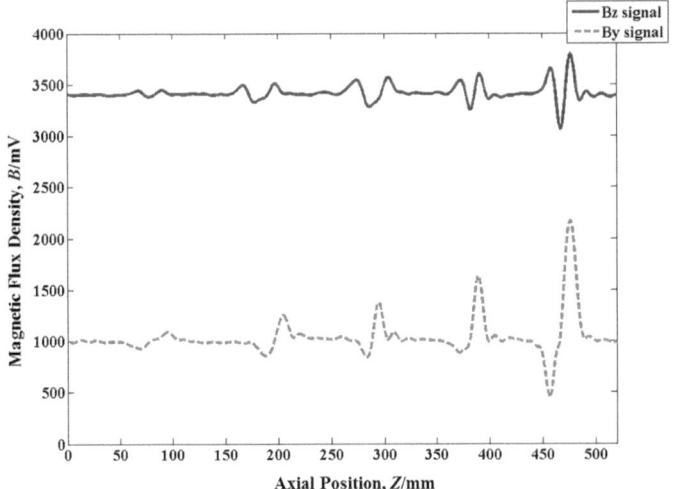

Fig. 9 Bz and By signals

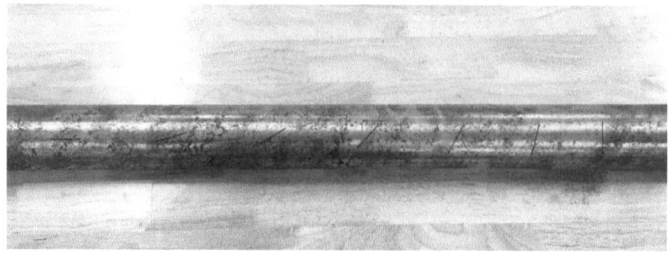

Fig. 10 Pipeline with the cracks in the range of 0°–90° orientation

through the system in Fig. 7 at a speed of 10 mm/s. The data of the sensor array were obtained and drawn in the MATLAB. The C scan results are shown in Fig. 11. It can be seen from the results that the orientation of the cracks can be recognized clearly. Moreover, with the increasing of the crack orientation, the B_z signal varies from dip to peak gradually. The B_z and B_y signals of the crack of 0°, 15° and 45° orientation are caused by the electric current perturbation effect and the signals of the crack of 60°, 75° and 90° orientation are caused by the magnetic flux leakage effect. Comparing with the results in simulation, we can see that the numerical and experimental results have the similar variation law and mechanism that using the combination effect of electric current perturbation and magnetic flux leakage, the arbitrary orientation crack can be detected.

4   Conclusions

In this paper, a Helmholtz-coil probe has been proposed to detect oriented cracks on the surface of the pipeline. A FEM model of this probe is put forward through the COMSOL. Based on the model, the relationship between the crack orientation and Bz and By signal is analysis. Finally, the structure of the Helmholtz-coil probe with a sensor array is designed and the experimental studies are carried out to test the detectability of oriented crack. The results of the simulations and experiments show that: using the combination effect of the current perturbation and magnetic flux leakage, the oriented crack on the surface of pipeline can be recognized clearly. The length of the cracks can be measured through the By signals.

Although, the detection and length measurement of the oriented cracks through the probe in this paper have been proved, the depth of the oriented crack can't be sized directly from the Bz signals. In the future work, we will focus on the depth sizing algorithm of oriented cracks through the Helmholtz-coil probe. Moreover, in the experiment, the artificial cracks were detected. In the next stage, some actual cracks or closed SCC will be tested.

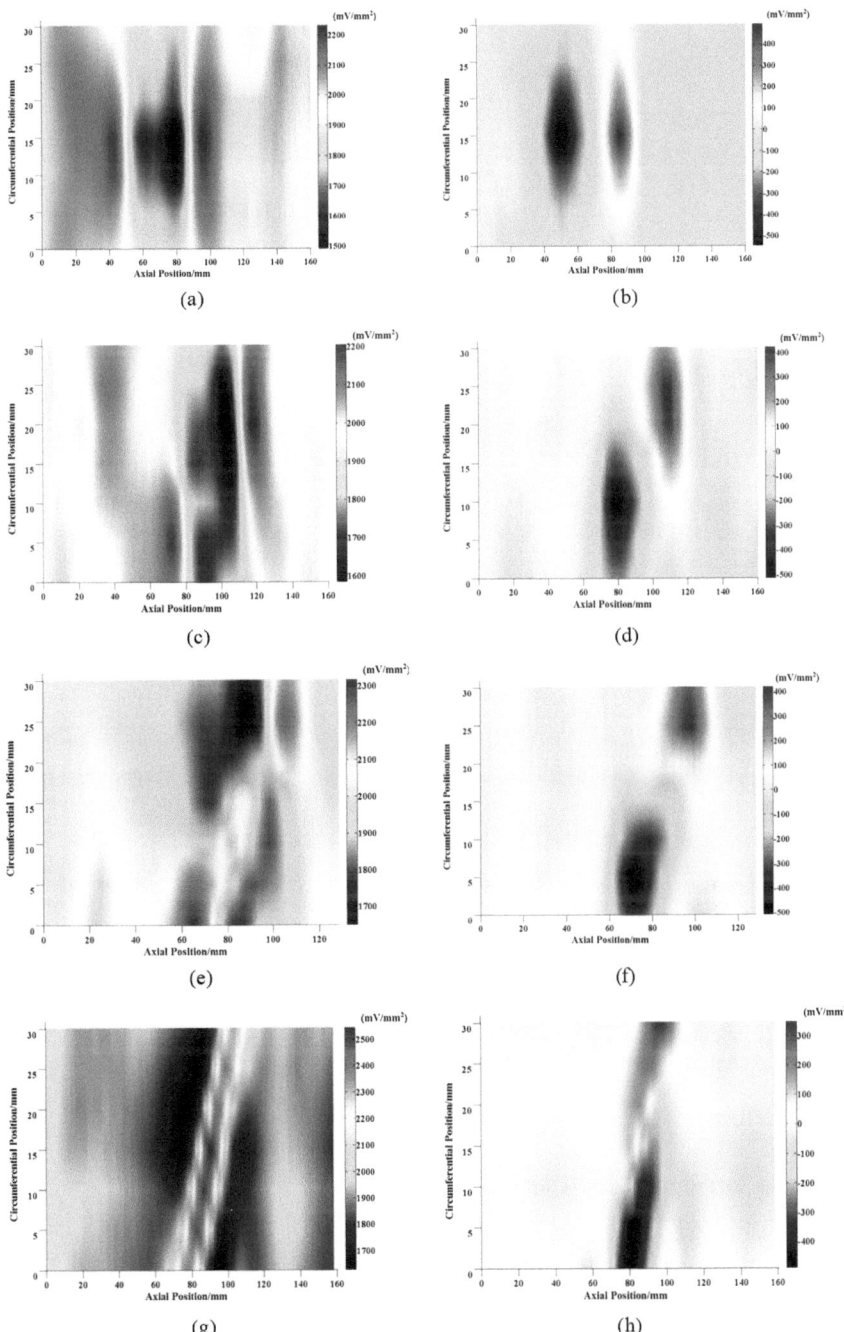

Fig. 11 C scan results of the oriented crack ranging from 0° to 90° with 15° interval, **a, c, e, g, i, k, m** are the Bz signals, **b, d, f, h, j, l, n** are the By signals

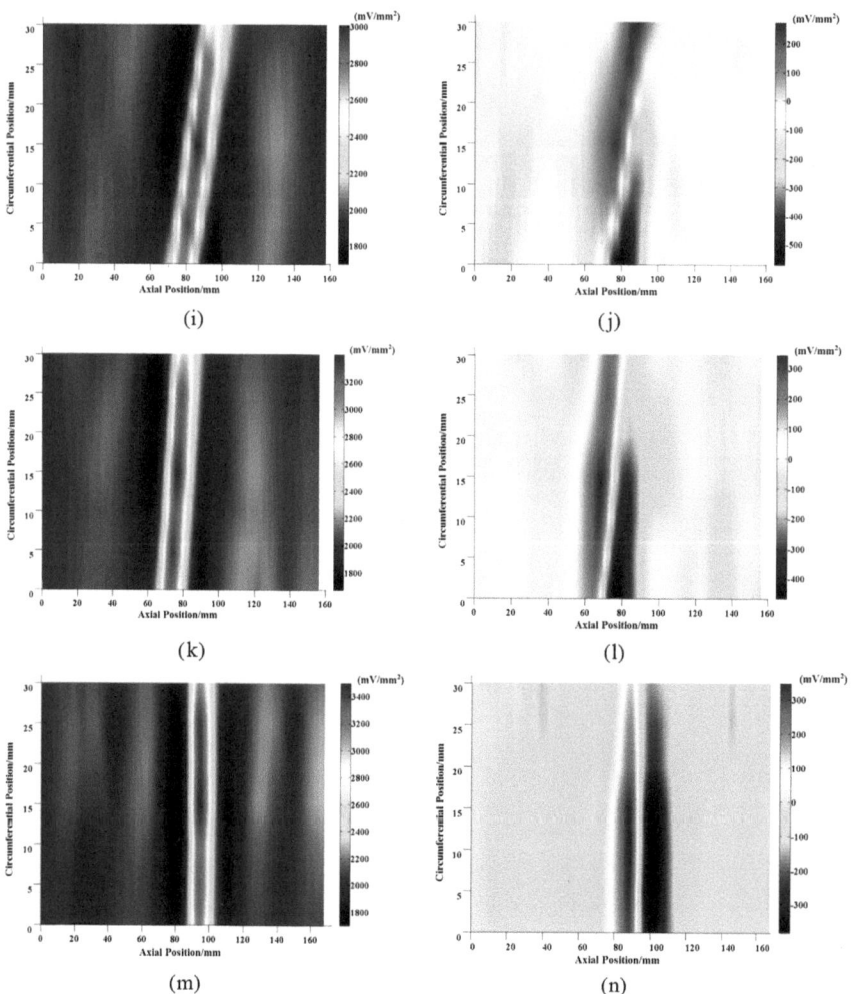

Fig. 11 (continued)

References

1. Cheng YF (2013) Stress corrosion cracking of pipelines
2. Yusa N, Hashizume H, Urayama R et al (2014) NDT&E Int 61:29–34
3. Beller M (2015) Pipeline Gas J 242(11):30–36
4. Mak DK (1985) Ultrasonics 23(5):223–226
5. Smith M, Sutherby R (2005) Insight: non-destructive testing & condition monitoring. 47(12):765–768
6. Jarvis R, Cawley P, Nagy PB (2016) NDT&E Int 81:46–59
7. Wu J, Sun Y, Kang Y et al (2015) IEEE Trans Magn 51(1):1–7
8. Li W, Yuan X, Chen G et al (2014) NDT&E Int 67(8):17–23
9. Trout SR (1988) IEEE Trans Magn 24(4):2108–2111

10. Hatsukade Y et al (2007) IEEE Trans Appl Supercond 17(2):780–783
11. Wu B, Wang Y J, Liu X C et al (2015) Smart Mater Struct 24(7)

Circumferential Current Field Testing System with TMR Sensor Array for Non-contact Detection and Estimation of Cracks on Power Plant Piping

1 Introduction

Piping is a critical component in power plant, such as the steam piping, hot reheat steam piping, cold reheat steam piping, feed water piping, cooling water piping and auxiliary piping. The piping serves in a harsh environment due to the noticeable cyclical pressure and thermal stress [1–3]. The pressure can reach 30 MPa and the temperature is more than 500 °C inside the piping. The high pressure and temperature can cause initial thermal fatigue cracks on the outer surface of piping [4, 5]. What's more, the initial thermal fatigue cracks gather and grow quickly by the cyclical stress, which cause the leakage and failure of the piping [6, 7]. Therefore, it is a critical issue to detect and estimate cracks on power plant piping before the failure happens [8, 9].

There are various non-destructive testing (NDT) methods for the detection of cracks on piping. Because of the closed piping system, the inner detection techniques, such as pipeline pigs, cannot be applied for the detection of cracks on power plant piping during in-service time. The ultrasonic testing (UT) needs coupling medium, which makes it an inadequate method in high temperature environment [10]. The infrared testing is a good method for surface cracks detection. However, the temperature changes constantly on the surface of piping. Thus it is hard to perform the infrared testing on power plant piping [11]. The magnetic flux leakage testing (MFL) is not sensitive to the narrow thermal fatigue cracks because of the little leakage of magnetic field on stainless steel piping [12]. The eddy current testing (ET) is sensitive to lift-off, which cannot penetrate thick coatings [13]. The microwave waveguide imaging technique is mainly used for corrosion inspection on plate [14]. What's more, most of the methods as mentioned above need multiple scans to achieve a full 360° inspection of the piping surface.

The magnetic particle testing (MT) and penetrant testing (PT) are traditional NDT methods for surface cracks detection on power plant piping. However, most processes of MT and PT needs manual operation in the harsh environment. These

© The Author(s) 2025
W. Li et al., *Alternating Current Field Measurement Technique for Detection and Measurement of Cracks in Structures*, https://doi.org/10.1007/978-981-97-7255-1_9

methods just give surface information of the crack but no depth information which is more important to estimate the residual life of the piping. What's more, the coatings need to be removed or the attachments on the piping should be cleaned before NDT operation. The magnetic particles and penetrants should be cleaned and the coatings should be re-painted after the inspection, which is time-consuming and costly [15].

The current field perturbation NDT methods, such as alternating current field measurement (ACFM), alternating current potential drop (ACPD) technique, are promising techniques to inspect defects on conductive material [16, 17]. For the detection of cracks on piping, the coaxial solenoid loaded with sinusoidal excitation signal can induces the current field on the surface of piping in circumferential direction. In the center of the coaxial solenoid, the induced circumferential current field on piping can be regard as an approximate uniform field. When the crack is presented, the circumferential current field will be disturbed. The space magnetic field can be picked up by sensors around the piping without contact [18, 19]. The defects can be detected and evaluated by the distorted magnetic field.

The circumferential uniform current field testing system is presented for non-contact detection and estimation of surface cracks on the power plant piping in this paper. The circumferential current is induced on the surface of piping by a coaxial encircling excitation coil. The inducing frequency is optimized to balance the penetration depth and detection sensitivity. The circumferential uniform current field testing system can cover a full 360° area on the surface of piping by TMR sensor array in a one pass scan. All the cracks on the surface of piping can be identified and evaluated visually and efficiently by the space magnetic field without contact using the circumferential uniform current field testing system.

This paper is organized as follows. In Sect. 2, the finite element method (FEM) model is set up and the inducing frequency is optimized by the simulation model. In Sect. 3, the circumferential uniform current field testing system is developed and the crack detection experiments are carried out. In Sect. 4, the cracks are estimated by the distorted space magnetic field. In Sect. 5, the conclusion and further work are delivered.

2 Circumferential Current Field FEM Model

2.1 Model Set up

The 3D FEM model of the circumferential current field is set up by ANSYS software, as shown in Fig. 1a. The model consists of excitation coil, piping and axial crack. The excitation coil is coaxial with the piping. The axial crack lies on the outer surface of piping symmetrically. To simulate the detection process, a dynamic FEM model is developed. The excitation coil is driven at 0.1 mm step-size to scan along the piping. The material of excitation coil is copper and the piping is stainless steel. As shown in Fig. 1b, the lift-off of the excitation coil is 10 mm and the lift-off of sensors is 7 mm, which gives enough space for the coatings. The size of the crack is 20 mm in length, 0.3 mm in width, 2 mm in depth. A sinusoidal excitation signal (frequency 20 kHz, amplitude 2 V) is loaded on the excitation coil.

When the excitation coil is above the crack, the current field around the crack is extracted on the surface of pipe string. As shown in the bottom right corner of Fig. 1a, the circumferential current field is uniform when the crack is not presented. The uniform current field gathers and turns around at the tips of the axial crack. The disturbed uniform current field perturbs the space magnetic field above the crack [20].

Fig. 1 3D FEM model of circumferential current field testing probe. **a** FEM model. **b** Size of the FEM model

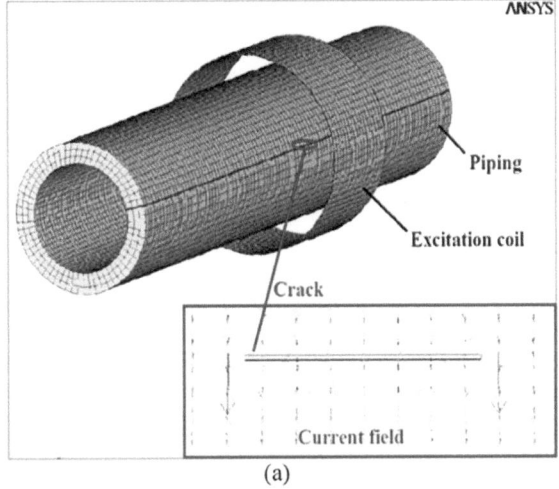

(a)

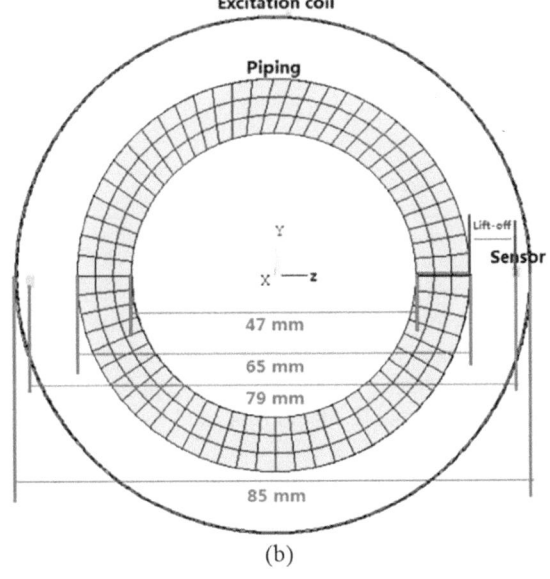

(b)

Fig. 2 *Bx* and *Bz*

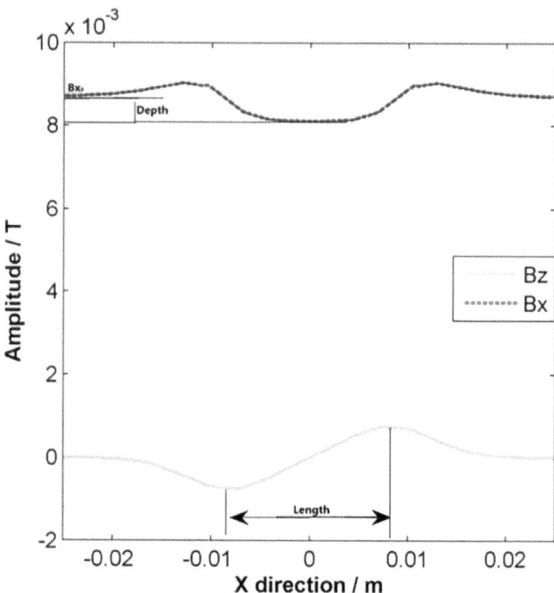

To get the distorted magnetic field caused by disturbed uniform current field, the magnetic field in axial direction (*Bx*) and radial direction (*Bz*) are picked up in the center of the excitation coil at the lift-off of 7 mm. Thus, as the excitation coil moves one step, the *Bx* and *Bz* are plotted once at each position, as shown in Fig. 2. The *Bx* shows a trough in the center of the crack due to the decrease of current density in the depth of the crack, which contains the depth information. Meanwhile, the *Bz* plots a positive and negative peak at the tips of the crack. The distance between the two peaks reflects the length of the crack.

2.2 Frequency Optimization

Due to the skin effect, inducing frequency determines the penetration depth on piping, which has a great significant influence on detection sensitivity of crack depth [21, 22]. The skin thickness is given in Eq. (1), where u_r is the relative magnetic permeability, u_0 is the magnetic permeability of free space, and σ is the electrical conductance, f is the inducing frequency. When applying a high frequency excitation signal on the excitation coil, the current field tends to concentrate in a thin layer flowing in circumferential direction, which goes against sizing crack depth [23]. When the inducing frequency is low, the current field has a large penetration depth, which is good for sizing crack depth. However, when the inducing frequency is too low, the induced current density is very weak on the surface of piping. Thus, the characteristic signals of crack are weak, which can be covered by noise easily. Therefore, it is necessary to optimize the inducing frequency to balance the detection sensitivity and penetration depth [24].

$$\delta = 1 / \left(\pi u_r u_0 \sigma f\right)^{1/2} \tag{1}$$

As mentioned above, the Bx contains the depth information of crack. Hence, the effect of inducing frequency on Bx is analyzed using the FEM model. The background magnetic field is extracted at different frequencies (50 Hz–60 kHz), as shown in Fig. 3a. The background magnetic field rises steeply as the frequency increases from 50 Hz to 20 kHz and becomes a slow climb after 20 kHz. It suggests that when the inducing frequency reaches 20 kHz, the current field saturates nearly on the surface of piping. So to get a deeper penetration depth, the inducing frequency should be lower.

The Bx with different inducing frequency is simulated, as shown in Fig. 3b. Initially the troughs of Bx are flat and becomes obvious after 1 kHz. The sensitivity of Bx is a critical parameter for the estimation of crack depth, which is defined as follows [25].

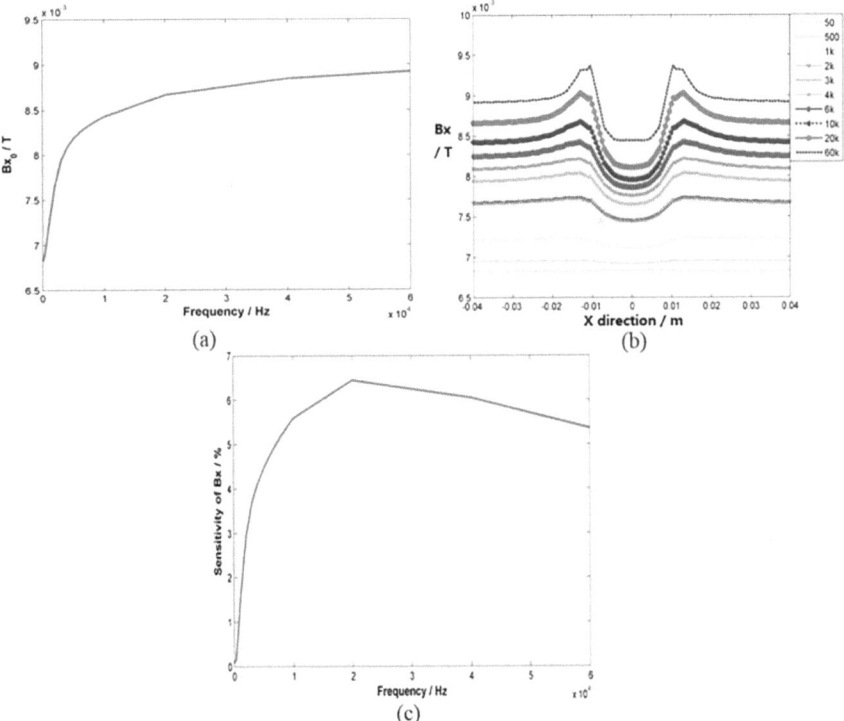

Fig. 3 Frequency optimization. **a** Background of Bx with different frequencies. **b** Bx with different frequencies. **c** Sensitivity of Bx with different frequencies

$$S_{Bx} = Bx_{\max}/Bx_0 \tag{2}$$

where $\Delta Bx_{\max}$ is the maximum distortion of Bx, Bx_0 is the background of Bx.

A high sensitive Bx signal will greatly improve signal to noise ratio and carry more crack depth information. As shown in Fig. 3c, the sensitivity of Bx increases steeply initially and reaches top at 20 kHz and then drops at last. Above all, the simulation results suggest that 20 kHz is the appropriate inducing frequency for high sensitive detection of crack depth.

3   Circumferential Uniform Current Field Testing System

3.1   Testing System

In practice, the piping is fixed and the probe is driven by mechanical devices or robots to scan the piping. In the laboratory, for convenience the probe is fixed on the scanner and the piping is driven to pass through the probe. The circumferential uniform current field testing system includes probe, excitation module, signal processing module, acquisition card, scanner and personal computer (PC), as shown in Fig. 4. The scanner is controlled by the programmable logic controller (PLC) in control cabinet. The piping is driven by the scanner to pass through the probe at a constant speed. The displacement information of the piping is recorded by the encoder. The probe induces a uniform current field on the surface of piping by the excitation coil. When a crack is presented, the uniform current field will be disturbed. The disturbed current field perturbs the space magnetic field. The magnetic sensors on the probe pick up the distorted magnetic field which is sent to signal processing box for signal processing and acquisition. The software in the computer shows the imaging of space magnetic field visually and estimates the size of the crack.

The probe consists of sensor module array, excitation coil and detachable nylon yoke, as shown in Fig. 5. The sensor module is made of two high-precision tunnel magneto resistance (TMR) magnetic sensors, whose operating temperature can reach 125 °C [26–28]. The two TMR sensors are sealed on each side of one common printed circuit board (PCB) and the sensitive axis of the sensors is orthogonal, which is used to measure the Bx and Bz. On the PCB, there are two primary amplifier chips (AD620) for Bx and Bz amplification (Bx 5 times and Bz 10 times). To achieve a full 360°detection of the piping surface, the sensor modules are installed on the yoke with an equal space as sensor array. In this paper, the sensor array is with 20° to each other in the circumferential direction of the piping [29]. There are 5 close TMR sensor arrays in the probe. The space magnetic field can be measured visually for imaging under the sensor array [30]. To avoid removing the coatings on the piping, the lift-off of sensor modules is 7 mm. Because the piping is closed, the yoke is separated in half symmetrically. Thus the yoke and sensors can be disassembled and installed to avoid the elbow pipes and flanges. When the two part yokes are encapsulated, the

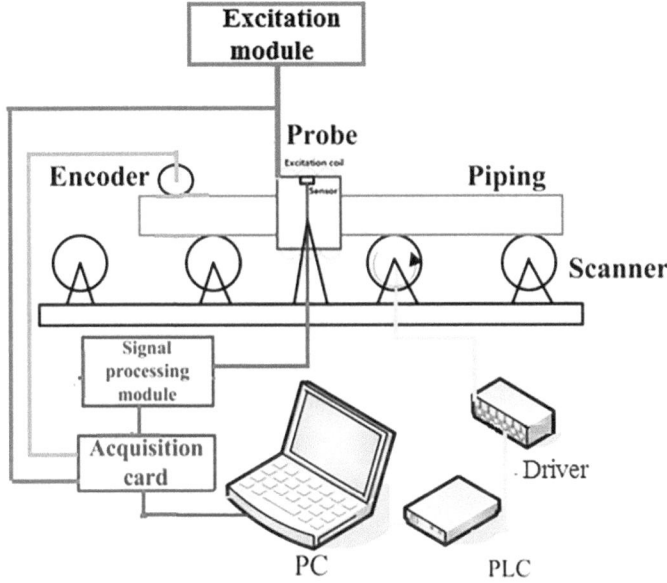

Fig. 4 Block diagram of circumferential uniform current field test system

excitation coil (copper wire whose diameter is 1 mm) is wound on the yoke with 50 turns.

There are four main parts: power system, excitation module, signal processing module and acquisition card in the signal processing box. The power system is a rechargeable lithium ion battery which provides power for each module. The excitation model transmits sinusoidal signal with 20 kHz frequency and 2 V amplitude, as shown in Fig. 6a. The signal processing module includes second amplifying circuit (Bx 10 times and Bz 10 times), band-pass filtering (10–30 kHz) and zeroing circuit, as shown in Fig. 6b. Thus the Bx and Bz is amplified 50 times and 100 times respectively as a whole. The Bx and Bz are filtered by the band-pass filtering and then calibrated by the zeroing circuit to keep the same zero point and scale. The Bx and Bz are captured by the acquisition card and then sent to the PC. The circumferential uniform current field testing system is set up, as shown in Fig. 6c.

4 Experiments and Discussions

4.1 Crack Depth Estimation

The samples are two stainless steel pipes (external diameter 65 mm, inner diameter 47), as shown in Fig. 7. On No.1 sample, the length of cracks is the same (30 mm) and the depth is different (2, 4, 6, 8 mm and a through crack). On No. 2 sample, the

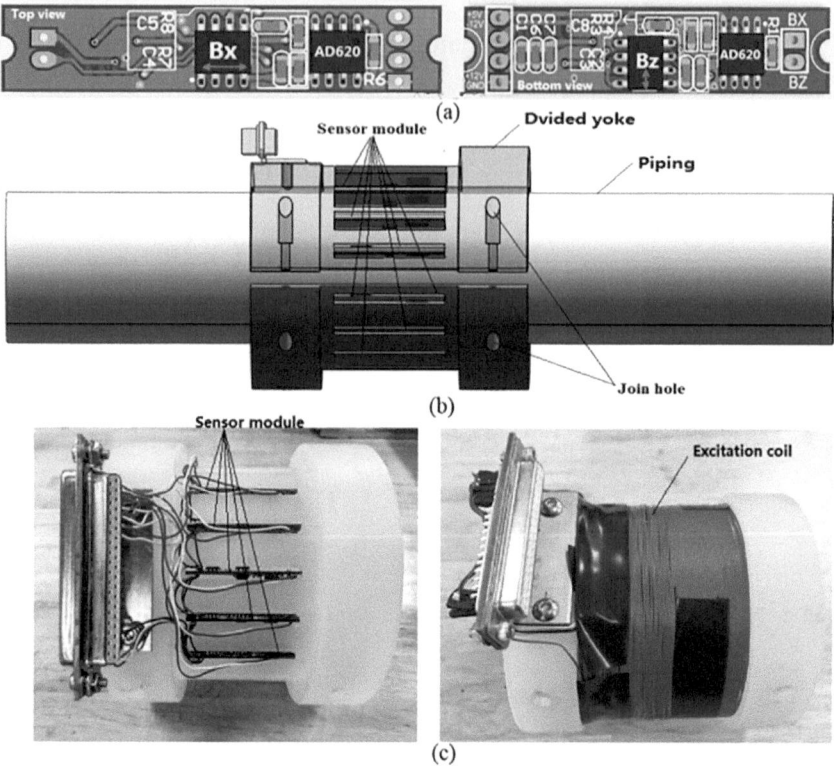

Fig. 5 Probe of circumferential uniform current field testing system **a** TMR sensor module. **b** Dvided yokes and sensors. **c** Photo of probe

length of cracks is different (55, 50, 45, 40, 35, 30 mm) and the depth is the same (4 mm). The width of all cracks is 0.3 mm.

The first sample (No. 1 simple) is driven by the scanner to pass through the probe at the speed of 10 mm/s (from 2 mm depth crack to through crack). The space magnetic field Bx around the crack on the No. 1 sample is shown in PC, as shown in Fig. 8. There are five troughs in Bx, while there are five opposite peaks in Bz at the same location. The characteristic of the Bx and Bz are in accord with the simulation results. The Bx and Bz are highlighted in the background magnetic field at the lift-off of 7 mm, which helps to recognize the cracks easily without contact. Especially, the maximum distortion values of Bx troughs and Bz peaks both appear in channel 4. Thus the location of the cracks is confirmed on piping according to the position of sensors in channel 4 o in the probe. Because the background of Bz approximates to zero, the distortion of Bz is more smooth and outstanding than Bx.

The Bx and Bz are selected in channel 4, as shown in Fig. 9. The troughs of Bx increase as the crack depth grows and the peaks of Bz goes up at the same time.

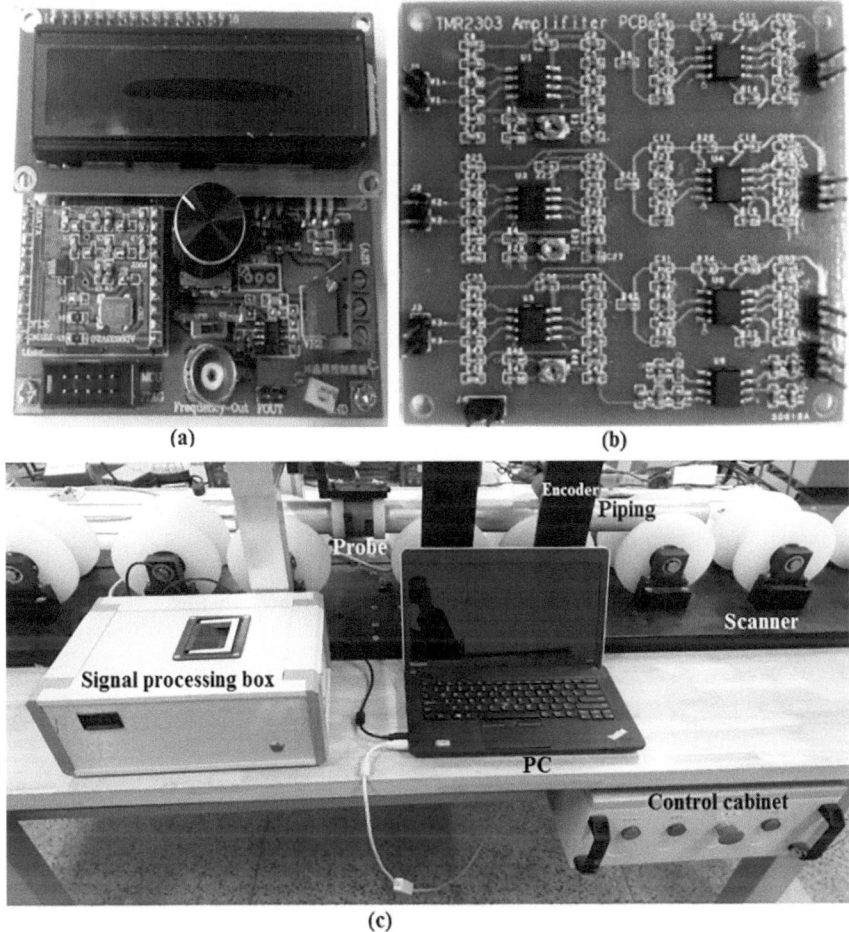

Fig. 6 Circumferential uniform current field testing system. **a** Excitation module. **b** Signal processing module. **c** Photo of testing system

The S_{Bx} with different depth cracks are shown in Fig. 10a. The maximum distortion values of Bz (ΔBz) with different depth cracks are shown in Fig. 10b.

The S_{Bx} and ΔBz almost increase linearly as the crack depth grows from 2 to 8 mm. However, there is a distortion point at the through crack. This is due to the material discontinuity in the through crack and the uniform current could not flow at the bottom of the crack. The current density drops sharply in the X direction and gathers more seriously at the tips of the crack.

Because there is a so well linear relationship between S_{Bx} and the crack depth (and the similar liner relationship between ΔBz and the crack depth), the calibration method is proposed to estimate the crack depth. The first two shallower cracks are

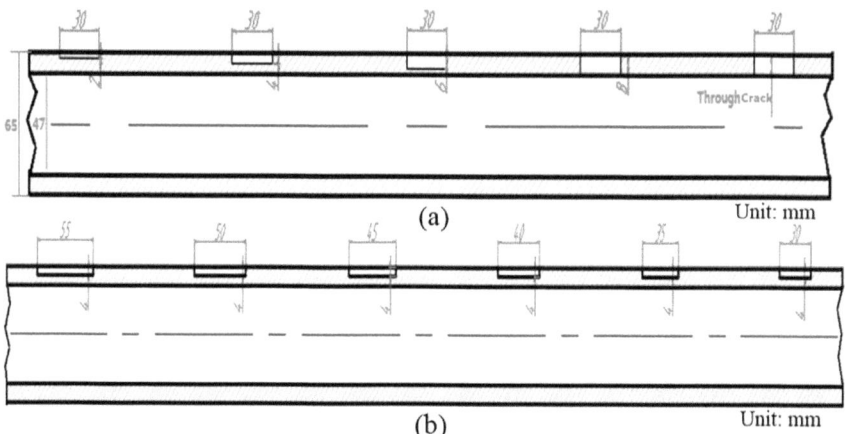

Fig. 7 Samples. **a** No. 1 sample with different depth cracks. **b** No. 2 sample with different length cracks

set as the calibrated crack to predict the last three cracks. According to the first two cracks, the depth of the other three cracks can be calibrated by Eqs. (3) and (4) fitted by Fig. 10a, b respectively:

$$D_s = 119S_{Bx} - 0.2432 \tag{3}$$

$$D_\Delta = 0.070098\Delta Bz + 0.8661 \tag{4}$$

where D_s is the crack depth estimated by S_{Bx}, D_Δ is the crack depth estimated by ΔBz.

Thus the last three cracks can be estimated by Eqs. (3) and (4). The predicted depths (PD) and relative errors are shown in Table 1. The 6 mm depth crack can be estimated by S_{Bx} and ΔBz. The relative errors (RE) are 13.0% and 2.6% respectively. The 8 mm depth and through crack are regarded as a break according the estimation by S_{Bx}. The 8 mm depth is estimated by ΔBz and the relative error is 8.6%. The through crack is identified as a break by S_{Bx} and ΔBz. If the first three cracks are set as calibrated crack, the relative errors of 8 mm depth crack are both less than 5% by S_{Bx} and ΔBz. We can make a conclusion that the crack depth can be estimated by S_{Bx} and ΔBz using the calibration method. Thus the residual thickness of power plant piping can be evaluated by periodic detection using circumferential uniform current field testing system.

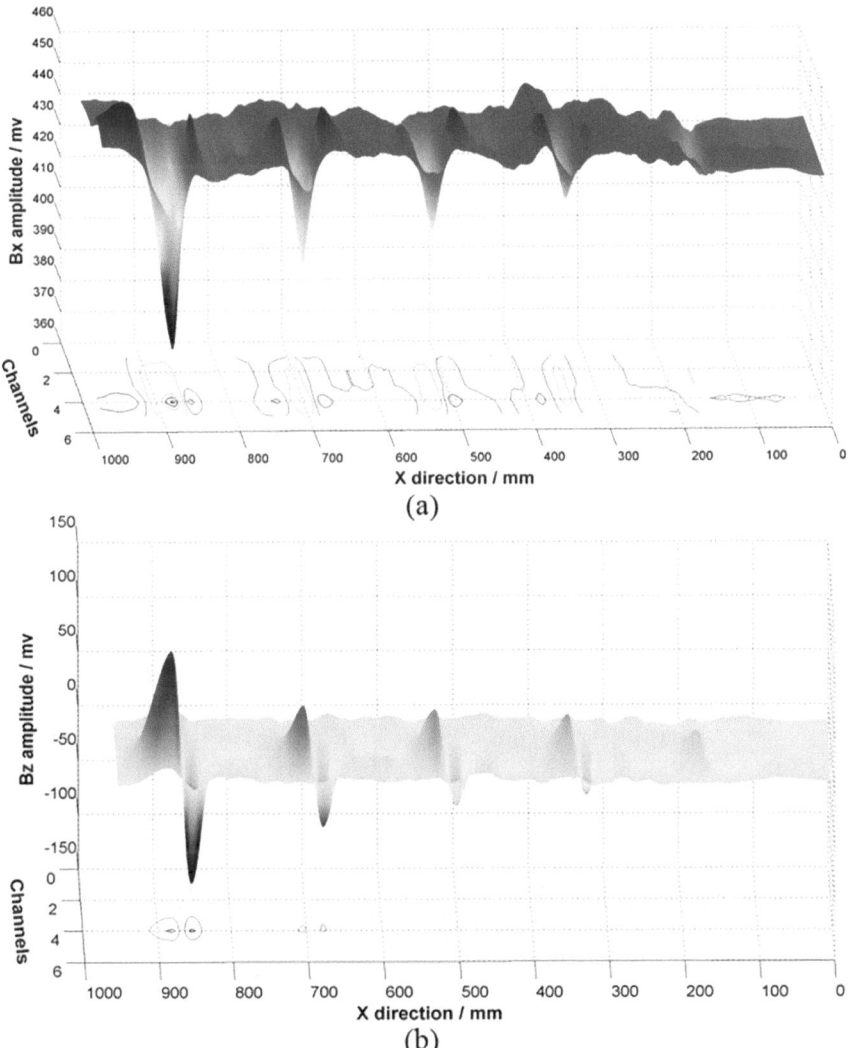

Fig. 8 Space magnetic field Bx and Bz of cracks on no. 1 sample

4.2 Crack Length Estimation

The second sample (No. 2 sample) is driven to pass through the probe at the same speed (from 55 mm length crack to 30 mm length crack). As shown in Fig. 11, the space magnetic field around the cracks on the No. 2 sample is plotted. There are 6 troughs in Bx and 6 peaks in Bz at the same location. Because the depth of the cracks on No.2 sample is the same, there are no obvious changes in the troughs of Bx and the peaks of Bz. Similarly, the troughs and peaks are located at channel 4.

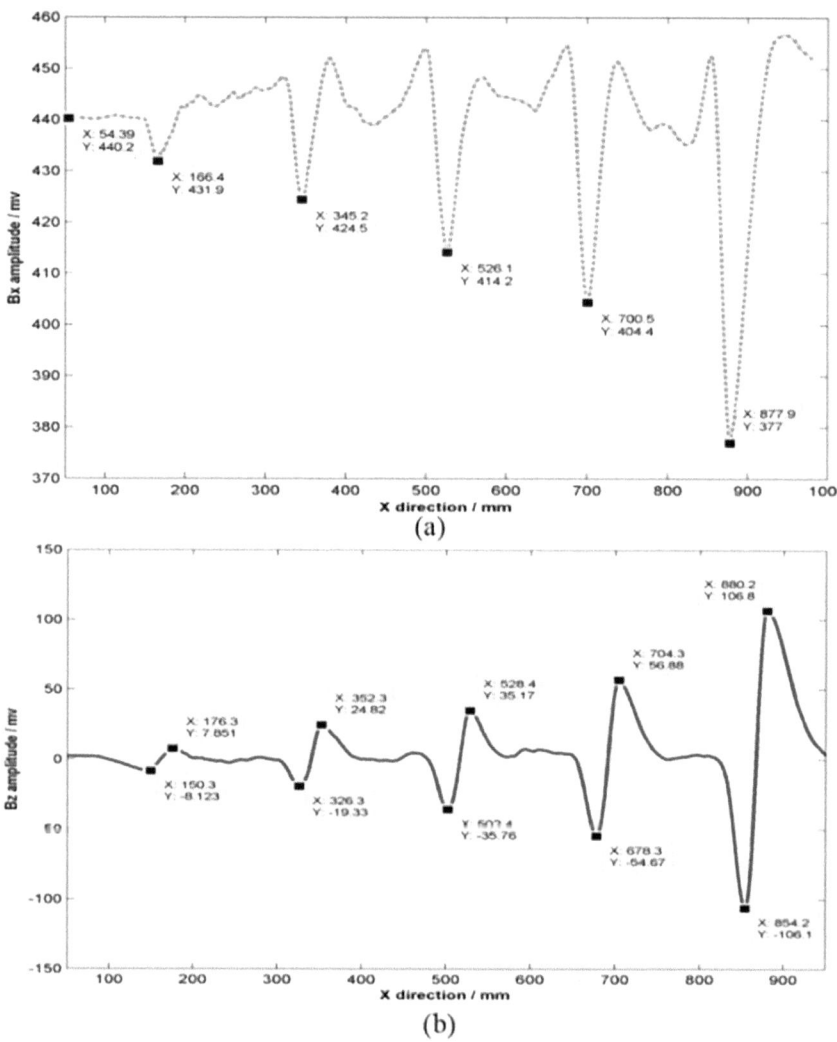

Fig. 9 Signals Bx and Bz in channel 4 of no. 1 samples. **a** *Bx*. **b** *Bz*

The *Bx* and *Bz* are selected in channel 4 and the encoder records the displacement information in the X coordinate, as shown in Fig. 12. Because the peaks of *Bz* locate at the tips of crack, the distance between the two opposite peaks of *Bz* (ΔL) reflects the length of cracks. As shown in Fig. 12b, the ΔL becomes narrow as the crack length decreases. However, the current field deflects clockwise at one tip of the crack and deflects anticlockwise at the other tip of the crack. The maximum magnetic field locates in the center of the deflected current field, which makes the ΔL less than the crack length.

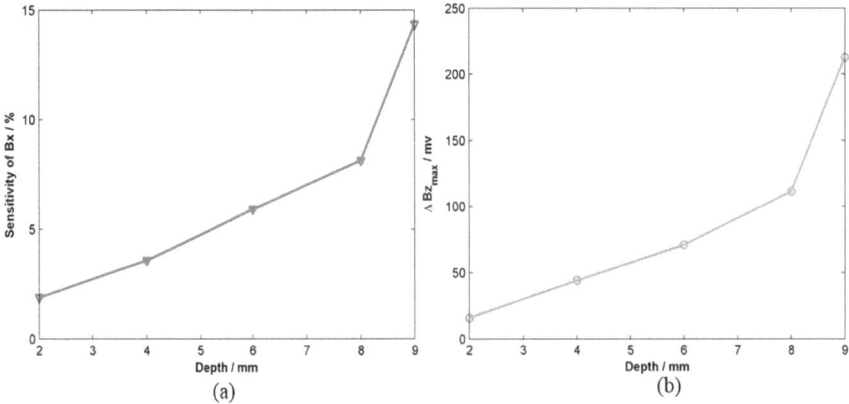

Fig. 10 **a** S_{Bx} with different depth cracks. **b** ΔBz with different depth cracks

Table 1 Estimated results of crack depth

Depth/ mm	2	4	6	8	Through
S_{Bx}/%	1.886	3.567	5.906	8.133	14.357
PD/mm	–	–	6.78	9.44	16.84
RE/%	–	–	13.0%	Break	Break
ΔBz/mv	15.974	44.150	70.930	111.550	212.900
PD/mm	–	–	5.84	8.69	15.79
RE/%	–	–	2.6%	8.6%	Break

The ΔL with different length cracks is shown in Fig. 13. ΔL goes up linearly as the crack length grows. ΔL is less than the actual length of the crack and the relative errors become unacceptable for short crack, as shown in Table 2. The calibration method is also proposed to estimate the length of the cracks. The first two cracks are set as calibrated crack to predict the last four cracks. The calibration equation of crack length (L) is given by Eq. (5) fitted by Fig. 13.

$$L = 1.042\Delta L + 2.917 \tag{5}$$

As shown in Table 2, the relative errors of the crack length drop sharply by the calibration equation. The calibrated crack length equals the actual length of the crack except the 45 mm length crack which has a tiny relative error. Thus the length of the crack on piping can be estimated by the distance between the peaks of Bz using the calibration method. What's more, the development of the crack length on power plant piping can be monitored and warned by periodic detection and estimation.

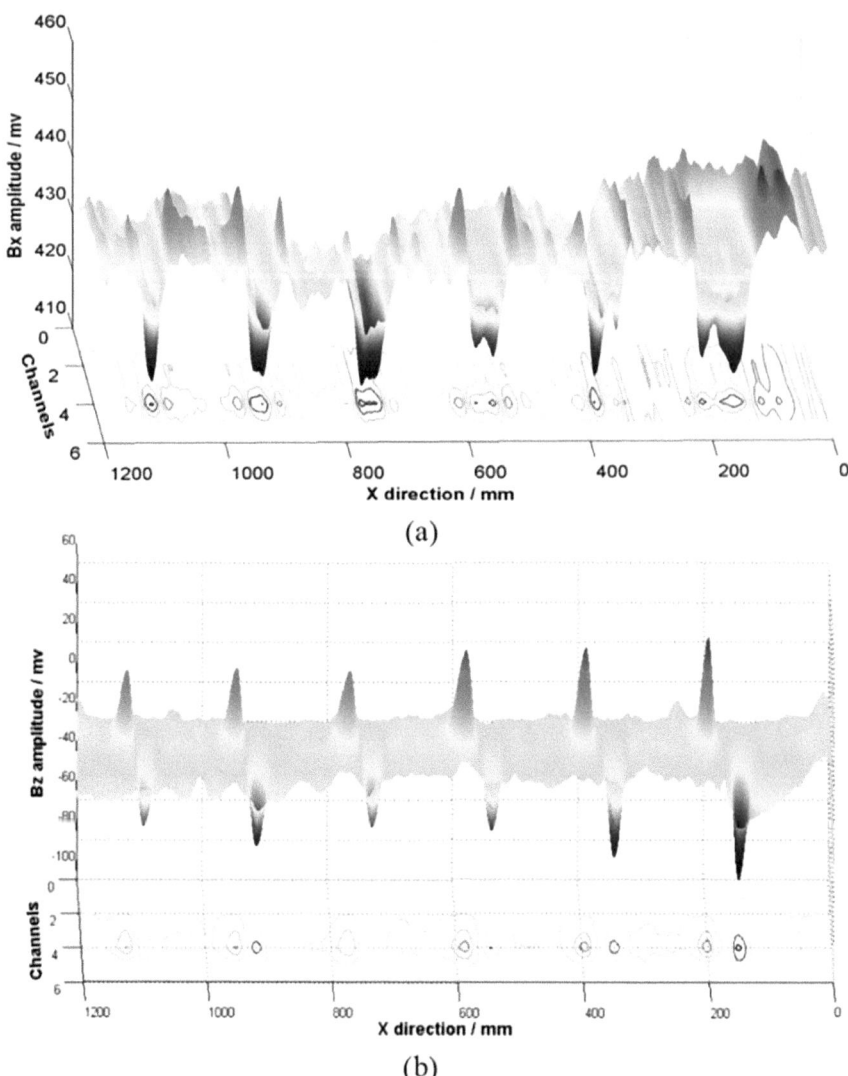

Fig. 11 Testing results of no. 2 sample

5 Conclusion and Further Work

This paper presents a novel circumferential uniform current field testing system for surface cracks detection and estimation on power plant piping. The dynamic FEM model is developed to extract the characteristic signals of the crack. The inducing frequency is optimized by the FEM model for sensitive and accurate detection of the crack. In the end, the circumferential uniform current field testing system is

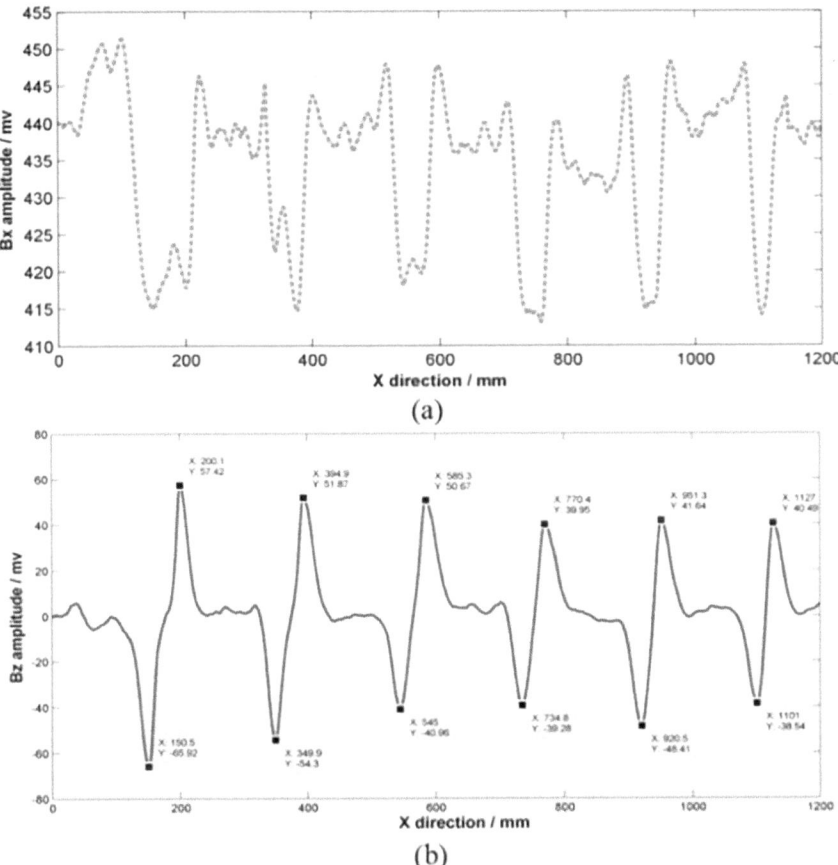

Fig. 12 Signals Bx and Bz in channel 4 of no. 2 samples

set up and the crack detection experiments are carried out. The results show that the cracks on piping can be detected and estimated visually and efficiently without contact by the circumferential uniform current field testing system with TMR sensor array using calibration method in a one pass scan. The depth of the crack can be estimated by the sensitivity of Bx and the maximum distortion of Bz. Meanwhile, the length of the crack can be evaluated by the distance between the opposite peaks of Bz. What's more, the circumferential uniform current field testing system provides a new method for non-contact, visual and efficient detection and estimation of cracks on power plant piping as an alternative technique to MT and PT. Further work will focus on monitoring the development of the crack using circumferential uniform current field testing system. Thus the propagation of the crack and the residual wall thickness can also be monitored and estimated.

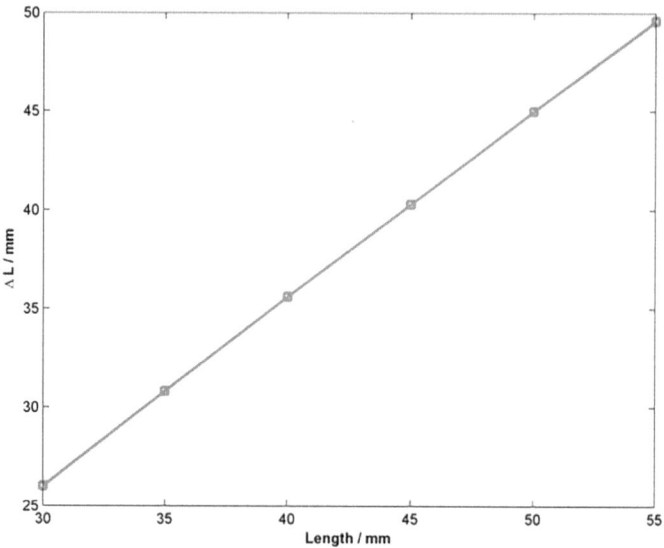

Fig. 13 ΔL with different length cracks

Table 2 Testing results of crack length

Length/mm	55	50	45	40	35	30
ΔL/ mm	49.6	45.0	40.3	35.6	30.8	26.0
RE/%	9.8	10.0	10.4	11.0	12.0	13.3
I_ι	–	–	44.9	40.0	35.0	30.0
RE/%	–	–	0.2	0	0	0

References

1. Jahromi AA, Firuzabad MF, Parvania M (2012) Optimized midterm preventive maintenance outage scheduling of thermal generating units. IEEE Trans Power Syst 27:1354–1365
2. Zhang X, Zhang JS, Gockenbach E (2015) Estimation of leak rate through circumferential cracks in pipes in nuclear power plants. Nucl Eng Technol 47:332–339
3. Amjady N, Daraeepour A (2011) Midterm demand prediction of electrical power systems using a new hybrid forecast technique. IEEE Trans Power Syst 26:755–765
4. Locatelli ML, Khazaka R, Diaham S, Pham CD, Bechara M, Dinculescu S, Bidan P (2014) Compact thermal failure model for devices subject to electrostatic discharge stresses. IEEE Trans Power Electron 29:2281–2288
5. Zhou YZ, Miao M, Salcedo JA, Hajjar JJ, Liou JJ (2015) Compact thermal failure model for devices subject to electrostatic discharge stresses. IEEE Trans Electron Devices 62:4128–4134
6. Saboonchi H, Ozevin D, Kabir M (2016) MEMS sensor fusion: Acoustic emission and strain. Sens Actuat A 247:566–578
7. Guedri A (2013) Reliability analysis of stainless steel piping using a single stress corrosion cracking damage parameter. Int J Press Vessels Pip 111:1–11

8. Gao YL, Tian GY, Li KJ, Ji J, Wang P, Wang HT (2015) Multiple cracks detection and visualization using magnetic flux leakage and eddy current pulsed thermography. Sens Actuat A 234:269–281

9. Huang H, Mawby PA (2013) A lifetime estimation technique for voltage source inverters. IEEE Trans Power Electron 28:4113–4119

10. Day P (2003) An automated ultrasonic inspection method for thickness checking of dryer cylinders for paper making machines. Insight 45:628–631

11. Altet J, Gomez D, Perpinyà X, Mateo D, González JL, Vellvehi M, Jordà X (2013) Efficiency determination of RF linear power amplifiers by steady-state temperature monitoring using built-in sensors. Sens Actuat A 192:49–57

12. Afzal M, Udpa S (2002) Advanced signal processing of magnetic flux leakage data obtained from seamless gas pipeline. NDT&E Int 35:449–457

13. Zhou DQ, Wang J, He YZ, Chen DW, Li K (2016) Influence of metallic shields on pulsed eddy current sensor for ferromagnetic materials defect detection. Sens Actuat A 248:162–172

14. Zhang H, He Y, Gao B, Tian GY, Xu L, Wu R (2016) Evaluation of atmospheric corrosion on coated steel using K-band sweep frequency microwave imaging. IEEE Sens J 16:3025–3033

15. Xie SJ, Chen ZM, Takagi T, Uchimoto T (2015) Quantitative non-destructive evaluation of wall thinning defect in double-layer pipe of nuclear power plants using pulsed ECT method. NDT&E Int 75:87–95

16. Li W, Yuan XA, Chen GM, Ge JH, Yin XK, Li KJ (2016) High sensitivity rotating alternating current field measurement for arbitrary-angle underwater cracks. NDT&E Int 79:123–131

17. Nativel EL, Talbert T, Martiré T, Joubert C, Daudé N, Falgayrettes P (2015) Near-field electromagnetic tomography applied to current density reconstruction in metallized capacitors. IEEE Trans Ind Inform 20:11–16

18. Zhu F, Zhang ZC, Luo JR, Zhang YW (2010) Investigation of the failure mechanism for an s-band pillbox output window applied in high-average-power klystrons. IEEE Trans Electron Devices 57:946–951

19. Qiu JZ, Sullivan CR (2013) High-frequency resistivity measurement method for multilayer soft magnetic films. IEEE Trans Power Electron 28:3581–3590

20. Li W, Chen GM, Yin XK, Zhang CR, Liu T (2013) Analysis of the lift-off effect of a U-shaped ACFM system. NDT&E Int 53:31–35

21. Fan MB, Wang Q, Cao BH, Ye B, Sunny AI, Tian GY (2016) Frequency optimization for enhancement of surface defect classification using the eddy current technique. Sensors 16:1–16

22. Sunny AI, Tian GY, Zhang J, Pal M (2016) Low frequency (LF) RFID sensors and selective transient feature extraction for corrosion characterisation. Sens Actuat A 241:34–43

23. Phan KL, Boeve H, Vanhelmont F, Ikkink T, Jong FD, Wilde HD (2016) Influence of metallic shields on pulsed eddy current sensor for ferromagnetic materials defect detection. Sens Actuat A 248:162–172

24. Brink EA, Hofsajer IW (2013) Analytical approach for determining the frequency dependent characteristics of multipath conductive structures. IEEE Trans Power Electron 29:5835–5845

25. Noroozi A, Hasanzadeh RPR, Ravan M (2013) A fuzzy learning approach for identification of arbitrary crack profiles using ACFM technique. IEEE Trans Magn 49:5016–5027

26. Arias SIR, Muñoz DR, Cardoso S, Ferreira R, Freitas PJPD (2015) Total ionizing dose (TID) evaluation of magnetic tunnel junction (MTJ) current sensors. Sens Actuat A 225:119–127

27. Dabek M, Wisniowski P (2015) Dynamic response of tunneling magneto resistance sensors to nanosecond current step. Sens Actuat A 232:148–150

28. Lim H, Lee S, Shin H (2015) Advanced circuit-Level model for temperature-sensitive read/write operation of a magnetic tunnel junction. IEEE Trans Electron Devices 62:666–672

29. Li W, Yuan XA, Chen GM, Yin XK, Ge JH (2015) A feed-through ACFM probe with sensor array for pipe string cracks inspection. NDT&E Int 67:17–23

30. Stefanov KD (2014) A statistical model for signal-dependent charge sharing in image sensors. IEEE Trans Electron Devices 61:110–115